European Working Environment in Figures

Availability and quality of occupational health and safety data in sixteen European countries

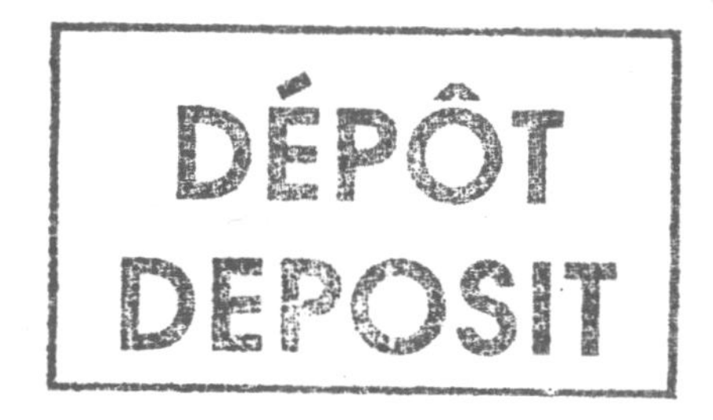

EF/96/12/EN

About the authors

This consolidated publication has been prepared for the European Foundation for the Improvement of Living and Working Conditions by three researchers from the Netherlands Institute for the Working Environment NIA, Amsterdam:

Sonja Nossent	Senior Researcher, specialised in topics concerning the working environment, in particular at sectorial level, both nationally and internationally.
Bert de Groot	Junior Researcher, skilled in quantitative studies on health and safety output and the working environment, both nationally and internationally.
Frans Verboon	Researcher, specialised in topics concerning health and safety output and social security, both nationally and internationally.

The final editing has been done by:

Sheila Pantry	Sheila Pantry Associates, Sheffield, United Kingdom Information Services Management Consultant, formerly Head of the Information Services of the United Kingdom Health and Safety Executive. Specialises in computerised services, creating CD-ROMs, training and setting up information services with a particular emphasis on all aspects of health and safety at work.

European Working Environment in Figures

Availability and quality of occupational health and safety data in sixteen European countries

Sonja Nossent, Bert de Groot, Frans Verboon
Netherlands Institute for the Working Environment NIA, Amsterdam

Sheila Pantry (final editing)

Sheila Pantry Associates, Information Services Consultancy,
Sheffield, United Kingdom

European Foundation
for the Improvement of Living and Working Conditions
Loughlinstown, Dublin 18, Ireland
Tel: +353 1 282 6888 Fax: +353 1 282 6456

Cataloguing data can be found at the end of this publication

Luxembourg: Office for Official Publications of the European Communities, 1996

ISBN 92-827-6552-0

Printed in Ireland

Preface

Information and data, which can pinpoint risk factors in the workplace, are a crucial starting point for the preparation of efficient measures to improve the working environment. To this end, most European countries have established monitoring systems. However, these systems differ in kind and were built to help policy formation at national rather than at European level.

To what extent are such national data available and valid? To what extent can they be utilized to describe the working environment at European level? What do the existing data sources tell us about common exposures and risks among workers across Europe? Where are the most urgent needs for further data gathering, quality improvement or co-ordination?

These are the main questions addressed by this publication which forms part of the Foundation's programme on monitoring the working environment in Europe.

This report has been prepared before the establishment of the European Agency for Safety and Health at Work; the objective of the Agency, as defined in Council Regulation (EC) N°2062/94, "shall be to provide the Community bodies, the Member States and those involved in the field with the technical, scientific and economic information of use in the field of safety and health at work".

We hope that this report will be useful for the Agency when it starts its work.

Henrik Litske is the Foundation's Research Manager in charge of this project.

Clive Purkiss
Director

Eric Verborgh
Deputy Director

Acknowledgements

The authors of this report, **Sonja Nossent, Bert de Groot** and **Frans Verboon** of the Netherlands Institute for the Working Environment NIA, Amsterdam, and the editor **Sheila Pantry** of Sheila Pantry Associates in the United Kingdom, have based their findings on a number of working papers covering five clusters of countries.

Marc de Greef et al. of the Association Nationale pour la Prévention des Accidents du Travail, ANPAT/NVVA in Brussels covered Austria, Belgium, France, Germany and Luxembourg.

Peter Laursen of the Danish Working Environment Service in Copenhagen covered Denmark, Finland, Norway and Sweden.

Maria Dolores Solé of the Instituto Nacional de Seguridad e Higiene en el Trabajo, INSHT in Barcelona covered Greece, Italy, Portugal and Spain.

Sonja Nossent, Bert de Groot and **Frans Verboon** of the Netherlands Institute for the Working Environment NIA in Amsterdam covered the Netherlands.

Sheila Pantry of Sheila Pantry Associates in the United Kingdom covered Ireland and the United Kingdom.

A large number of people have been contacted throughout Europe and they have contributed with data and information. These are too many to name here; they are listen in Annex 1.

A Steering Group was formed to support the project, which included the researchers, and further consisted of the following members:

Emilio Castejón	Instituto Nacional de Seguridad e Higiene en el Trabajo INSNT, Barcelona
Frits Claus	Ministry of Social Affairs and Employment, The Hague
Johnny Dyreborg	Eurostat, The European Commission, Luxembourg
Kurt Kreizberg	Bundesvereinigung der Deutschen Arbeitgeberverbände, Cologne
Karl Kuhn	Bundesanstalt für Arbeitsschutz, Dortmund
Tom Wall	Irish Congress of Trade Unions, Dublin

Thanks are due to the authors, the Steering Group and the large number of people who kindly contributed data and information and thereby made the project possible.

Henrik Litske
Research Manager

Contents

Page

Preface v

About the authors

Acknowledgements vii

Table of contents ix

Summary xi

1. Introduction 1

   1.1 Context of the project 1
   1.2 Aims of the project 4
   1.3 Structure of the report and remarks to the reader 5

2. Methodology of the study 7

   2.1 Theoretical monitoring model 7
   2.2 National data gathering 8
   2.3 Consolidation 11

3. The European labour market - main characteristics 15

4. Availability and quality of data on occupational health and safety in Europe 27

   4.1 Availability of data and sources of information 27
      4.1.1 Availability of data 27
      4.1.2 Sources of information 42
   4.2 Reliability and comparability of data 47
      4.2.1 Reliability of data 47
      4.2.2 Comparability of data 50
   4.3 Discussion and preliminary conclusions 54

5. Figures on occupational health and safety across Europe 59

- 5.1 Working environment 59
- 5.2 Health and safety output 72
- 5.3 Discussion and preliminary conclusions 88
  - 5.3.1 Working environment 88
  - 5.3.2 Health and safety output 91

6. Trends and strategies in Europe regarding data production 93

- 6.1 Identified problems and main topics in governments', social partners' and other authoritative organizations' policies 93
- 6.2 Developments and activities in countries regarding data gathering and production 96
  - 6.2.1 General developments 96
  - 6.2.2 Sectorial approach 98
- 6.3 Discussion and preliminary conclusions 100

7. Final conclusions and general discussion 103

- 7.1 The availability and quality of data 104
  - 7.1.1. Availibility of data 104
  - 7.1.2. Reliability of dat 108
  - 7.1.3. Comparability of data 109
  - 7.1.4. Information sources and organisations involved in data production 111
  - 7.1.5. Trends and developments in data production 112
  - 7.1.6. Final conclusions 114
- 7.2 The working environment and health and safety output 115

8. Recommendations for future policy 119

- 8.1 Recommendations on the working conditions across Europe 119
- 8.2 Recommendations on occupational health and safety monitoring across Europe 120

9. Bibliography 127

Annexes

1. Participating institutes and authors of national reports
2. Contents of the matrix and guidelines for national data gathering
3. ISCO-88 (COM): International Standard Classification of Occupation
4. NACE-1970: Statistical Classification of Economic Activity in the European Community
5. Exposure Classification System - a Proposal
6. Diagnostic Categories of Diseases
7. Overview on the total working population in 16 European countries
8. Age categories used in the matrix

Summary

European Working Environment in Figures: a tool for policy makers

Introduction

This new report by the European Foundation for the Improvement of Living and Working Conditions, Dublin shows that information is essential to pinpoint risk factors in the workplace, and is a crucial starting point for the preparation of efficient measures to improve the working environment. To this end, most European countries have established monitoring systems. However, these systems differ in kind and were built to help policy formation at national rather than at European level.

Because the report not only summarises monitoring data from various countries, but also provides an overview of some previous studies and initiatives in this field, the report presents a new and rather broad overview of the state of the art of monitoring health and safety in Europe. The report can thus be seen as another stepping stone in European health and safety policy.

This consolidated report is the result of the project 'the European Working Environment in Figures' in which information on sixteen European countries has been assembled. The report describes the aims and methodology of the project, the problems encountered, the results, and discussion and conclusions on the situation in Europe with respect to figures on occupational health and safety issues. It also contains recommendations for future activities to improve both the working conditions in Europe and the monitoring thereof. This summary highlights the headlines of these topics.

Context and aims of the project

The Foundation considered it useful to try to support the 'soft data' from the First European Survey on the Working Environment 1991-1992 (European Foundation, 1992) with 'soft and hard data' available from national monitoring systems in the (future) EU Member States, from the various Foundation networks and working groups, and from other Foundation projects and publications. Hence, this pilot project is closely linked with the other activities of the Foundation concerning occupational health and safety data gathering and harmonisation.

In the first place, this experimental study has been carried out to provide a quantitative overview of the working conditions in sixteen countries within Europe. Secondly, the aim was to compose an overview of possible gaps in the existing information on working conditions, by mapping out the availability, reliability and comparability of quantitative data on health and safety at the workplace. Furthermore, insight in the current developments regarding 'data production' in this field, was desired, both at national level and at European level. Hence, the specific aims of the study were formulated as follows.

1. To describe the working conditions in Europe based on existing data and information sources.
2. To identify common exposures and risks among workers across Europe, especially in relation to gender, age, occupation and economic sector.
3. To supplement the 'subjective data' from the First European Survey on the Working Environment 1991-1992 (the Eurosurvey) with 'objective' data from sixteen European countries.

4. To assess the extent of the lack of comparable data in Europe, and discuss the problems related to this.
5. To identify where further data gathering, quality improvement, or harmonisation is most urgent.

In summarising the gathered information, and trying to meet the objectives mentioned above, the consolidated report aims to comply with its main purpose: to provide a tool for policy makers, primarily at European level, but also at national level. It may enable them to discuss the situation with respect to occupational health and safety and the monitoring thereof in their country, and across Europe more fruitfully. Such discussions may further stimulate the development and implementation of future policies regarding data production, as well as policies on the improvement of the working conditions itself.

The report may however be of interest to a broader audience, such as statisticians, scientists, researchers and other experts in occupational health and safety. In other words, to anyone else interested in figures on the working environment and its output in various European countries.

Methodology of the study

To carry out the study, a network of researchers from national occupational health and safety organisations in five EU Member States was set up. A Steering Group was formed around these researchers, consisting of representatives from national governmental bodies, social partners, the Foundation, the European Commission and one of the research institutes involved.

For the purpose of the study, a concept of a 'theoretically ideal occupational health and safety monitoring system' was taken as a starting point, which defined three areas on which data should ideally be available: exposure to risk factors in the working environment (the input); interventions undertaken to prevent or reduce exposures (interventions); and consequences of these exposures, in terms of health and safety output such as occupational accidents and diseases, and in terms of ecnomic costs to society involved in occupational health and safety (the output). For several reasons (see section 2.1) it was decided not to collect data on interventions, nor on economic costs.

Hence, the study focussed on two monitoring areas, i.e. working environment and health and safety output, for which a so-called 'matrix' was developed along with explanatory guidelines. The matrix provided a uniform structure for the data gathering and reporting on each country. The researchers not only gathered information on their own country, but also on others for which they contacted relevant organisations in these countries. Thus, the following sixteen countries were covered in the study: Austria, Belgium, Denmark, France, Finland, Germany, Greece, Italy, Ireland, Luxembourg, Norway, the Netherlands, Portugal, Spain, Sweden and the United Kingdom.

The national data gathering mainly concerned quantitative data on the working environment (20 topics) and health and safety output (10 topics, 6 occupational and 4 general), and more qualitative information on trends and strategies regarding data production. As a general procedure, the national data had to be gathered from already existing national or international information sources, such as statistics, registers, surveys, other authoritative sources, and professional literature. In order to trace these information sources, researchers had access to the Foundation's HASTE Database, containing over 160 descriptions of information systems in EU Member States (European Foundation, 1995a), and to the 'Review of Surveys' (European Foundation, 1994f), which forms a catalogue of general, regional and sectoral surveys, amongst others in the sixteen countries studied.

Two other ways of data gathering were utilised. Eurostat provided data on most countries with respect to their national working populations and companies. Furthermore, researchers consulted national key informants from governmental bodies, unions, employer's organisations, and other authoritative organisations, such as centres for statistics. From them they obtained the information on trends and strategies in data production. Key informants also provided access to supplementary information and checked the draft national reports.

Although the researchers encountered various problems in the national data gathering, they succeeded in compiling sixteen national reports. These reports were all used in the consolidation, which was carried out by one of the participating research institutes. To ensure a correct presentation and interpretation of the data, two draft versions of the consolidated report were sent to all participants and members of the Steering Group, the first version for written comment, the last version for discussing it in an evaluative meeting.

Some of the problems encountered in the national data gathering also affected the consolidation. The main effect is, that in this report there is more emphasis on the availability and quality of data, rather than on the actual figures and conclusions on the European working environment and its health and safety output. This means that the objectives of the study mentioned earlier, are not equally met. The following sections will show that aims 1 and 2 have been met to some extent, aim 3 could not be met, whereas aims 4 and 5 have quite successfully been met.

The European labour market

The working population of the sixteen countries covered by this study amounts to circa 153 million people in total. The majority of this population is male. Almost 65% of this European working population is between 25 and 49 years old, some 20% is older than that (50-64 years of age, or 65 years and over), whereas circa 14% are young workers (15-24 years old).
The most common type of companies in the sixteen European countries is the category with 0-19 employees, ranging between 58-97% of all companies in the various countries. The majority of the workers however find employment in larger companies.
The largest economic sectors across the sixteen countries are 'Other services' and 'Distributive trade', employing respectively circa 39 and 26 million workers. The smallest sector, employing circa 2 million workers (1.5% of the total working population) concerns 'Energy and water'.

The European working environment in figures: main results and conclusions

Working environment

Based on the obtained figures on the European working environment it has been concluded that the working conditions in Europe can be described to some extent (aim 1) and that some common risk factors and risk groups across Europe can be identified (aim 2), but only indicatively. These conclusions are being elaborated hereafter.

Description European working environment

Addressing aim 1, it is concluded that the obtained data from existing national information sources enables descriptions of four aspects of the European working environment:

- the physical working environment with regard to: noise, vibrations, climate, and radiation;

- the psychosocial working conditions: working pace, job content, working hours, influence and control over work, and social interaction;
- physiological exposures: working postures and movements, and manual lifting;
- safety at work: overload/extension of the body, slips/trips/falls and other safety hazards.

Of the first three working environment aspects, the descriptions are almost exclusively based on workers' perceptions measured by questionnaire-based surveys from 6-10 countries. These descriptions only concern the amount of workers exposed, and not the severity, nor the duration of exposures since such data were not explictly requested in this study (yet obtained to some extent). The calculated weighted European averages of percentages of workers exposed to the various risk factors vary between ≥ 2% (radiation) and ≥ 51% (working pace). Due to cross-national differences in the surveys (see below under 'Monitoring occupational health and safety'), i.e. coverage of various sub-items under the risk factors, these figures can however only be taken as indicative.

The description of safety at work in Europe is based on occupational accidents registrations from 15 countries. The description consists of annual numbers of accidents in the EU (circa 8000 reported fatal accidents and circa 4.5 million reported accidents with work interruption), and of the most common causes of accidents in 11 countries (see below). Again, this description can only be taken as indicative, due to reliability and comparability problems (see below).

Descriptions of the chemical and biological aspects of the European working environment can not be made on data from national information sources, because not sufficient data was obtained. However, data from the Eurosurvey, held in 12 countries, do allow a description of chemical aspects of European working life to some extent, which concerns both percentages of workers exposed, as well as duration of exposures. This means that exposure to biological risk factors is the only aspect of working life which can not be described yet at European level.

Common risk factors and risks

In relation to aim 2 it has first of all been concluded that four risk factors seem most common across Europe, since exposure to these factors is reported over the largest proportion of the working population:

- defective job content;
- strenuous working pace;
- lack of influence and control over one's own work;
- strenuous working postures and movements.

These indications are derived from a combination of three kinds of analysis: two based on the material obtained in this study stemming from national information sources, and one on the outcomes of the Eurosurvey. Due to matters of reliability and non-comparability of data, the identification is indeed not more than indicative.

The same cautions had to be applied in identifying common risks across Europe which regard:

- main causes of occupational accidents:	* overload/extension of the body;
	* slips/trips/falls;
- main diagnoses of occupational diseases:	* skin diseases;
	* musculo-skeletal disorders;
	* hearing disorders;
- main diagnoses of general mortality:	* diseases of the circulatory system;
	* neoplasms.

Under reporting, cross-national differences in data producing methodologies and the non-direct work-relatedness of the data determine the indicative character of this common risks pinpointing.

Main risks and future risks
The indicatively identified common risk factors and common risks show overlap on one factor. It could therefore be concluded that 'strenuous working postures' and 'overload/extension of the body' form main hazards for workers across Europe, often resulting in musculo-skeletal disorders. Regarding future trends, key informants have put forward that the rise of new technology may pose new risk factors and that already a shift in focus can be observed from the 'traditional' physical hazards to the more 'modern' hazards, such as stress.

Common risk groups
With respect to common risk groups it has been concluded that, compared to female workers, male workers are more at risk to be exposed to noise and vibrations, and to be subject to fatal accidents and accidents with work interruption (higher rates). Furthermore, the (metal)manufacturing industries and building/engineering sectors are concluded to be relatively hazardous sectors across Europe in relation to occupational accidents. No risk groups could be identified among age groups and occupational groups, due to lack of sufficient, comparable data.

Monitoring occupational health and safety
The obtained data from national information sources have further led to three conclusions that regard the coverage and quality of occupational health and safety monitoring in Europe. Firstly, these conclusions are presented and after that further elaborated.

With respect to aim 3, it has been concluded that this aim can not be met with the material obtained. Supplementation of the subjective data from the Eurosurvey with objective or hard data is not feasible, firstly because objective working environment data, aggregated at national level, appeared to be too scarce. Secondly, it appeared that there is not really hard data on occupational health and safety output at European level (see also below).

In relation to aim 4, it has been possible to identify in detail which occupational health and safety topics in Europe suffer from lack of data, from unreliability of data, and from cross-national incomparability of data, whilst taking into account that the matrix used in this study in fact formed a filter in all existing data, and taking the 'theoretically ideal monitoring system' as a reference.

In relation to aim 5, the areas for which further data gathering, quality improvement or harmonisation is indicated, have been identified (see below where the topics suffering from unavailibity, unreliability and incomparability of data are being specified). It was further concluded that the determination of the most urgently needed improvements in data production, requires a discussion at European level (see further below under 'Policy options for the future').

Availability of data
Lack of data from national information sources has been concluded to prevail with respect to:
- 9 working environment risk factors: lighting, pressure, forms of payment, traumatic experiences, handling tools/equipment, chemical substances, biological hazards, safety hazards, and 'other hazards';
- 5 health and safety output topics: occupational sickness absence, occupational morbidity, occupational disablement, general morbidity and general disablement.

On these topics data were provided by only five countries or less. Furthermore, objective data on the working environment is generally lacking, yet known to exist, e.g. resulting from inspection and enforcement activities, but not aggregated to national or sectorial level. In addition, monitoring of undertaken interventions and economic costs involved in occupational health and safety seem not to be common in the various countries.

Reliability of data

The reliability of the data obtained has generally been concluded to be good with respect to validity (indeed measuring what is intended to be measured), representative value (covering entire national working populations) and up-to-date-ness (data mostly stemming from 1992-1993 or 1990-1991). On reproducability no conclusions were drawn and further analysis was felt needed to assess this quality criterion. Concluded to be problematic however, is the objectivity or 'hardness' of the data. Obtained working environment data is mainly subjective, stemming from questionnaire-based surveys, and not being confirmed by objective data from e.g. work place assessments or inspection and enforcement activities. Obtained health and safety output data suffer from significant under reporting of which the extent is however unknown, and from major cross-national differences in the data compiling methodologies (see below). In fact, only general mortality data seem hard, but these are not directly work-related.

Comparability of data

The cross-national comparability of the data obtained has been concluded to be the weakest aspect. Both for the working environment data, as well as the health and safety output data this deficit is due to different definitions, classification-categories and sub-items in the data reporting from the various countries. For example, occupational accidents are in one country registered after one day of sickness absence, whereas in others only after three days. The differences in classification-categories particularly concern age and occupation, and to a lesser extent economic sector. An illustration of the variety of sub-items under one risk factor is 'working pace', being measured as 'working at high speed', 'working under tight deadlines', 'working under pressure of time', and 'doing short repetitive tasks'. The working environment data moreover suffer to some extent from differences in the periods from which they stem. In fact, it appeared that only the obtained data on noise, fatal accidents, and general mortality are directly comparable. On the other topics comparison of data requires more in-depth analysis and probably conversion of categories and data.

Influential factors and consequences to European monitoring

Some influential factors behind the observed unavailability, unreliability and incomparability of data have been pinpointed and discussed. Besides the methodology of this study itself, these include historical factors and the (national) purposes for which data originally are produced. However, although these factors explain to some extent the observed deficits in the availability and quality of data, it has been concluded that 'the present European occupational health and safety monitoring system' suffers from:

- a limited ability to reliably pinpoint common or main risks and risk factors across Europe;
- a limited ability to identify common risk groups across Europe according to gender and economic sector, and inability to identify risk groups among age groups and occupational groups;
- a very limited ability to link causes, interventions and effects.

Furthermore, is was concluded that at the moment the best comparable working environment data available at European level, is the information from the Eurosurvey.

Information sources

Besides the conclusions addressing study aims 3-5, some additional conclusions have been drawn. Firstly, it has been concluded that the most common national information sources from which the national data stem, concern surveys (labour force; working environment), statistics (labour force; labour market; working environment; general), and reports (annual; research). The main providers of information respectively regard national bureaux of statistics, governmental bodies, national research institutes, social security organisations, and occupational insurance funds. Particularly the last two kinds of organisations and the governmental bodies play a far more important role in health and safety output data production, than in working environment data production.

Furthermore, it seems that national information sources containing national aggregated data on the working environment are entirely lacking in Greece, Portugal, Italy, Luxembourg, Ireland and the United Kingdom. For health and safety output data this seems to be the case for Norway and Portugal. However, this observation might be biased by the time restrictions of the project. In addition it was concluded that both at national level, as well as at European and global level more useful information sources exist than have been consulted in this study. To some extent these have been specified in the study in order to stimulate future use.

Trends and developments in data production
Additional conclusions further concern trends and developments in data production across Europe, which are based on assumed incomplete information from ten countries. In all these countries developments in (the improvement of) data production are taking place. In seven of them initiatives are deployed on both the working environment monitoring, as well as health and safety output monitoring, predominantly concerning integration. In the other three countries developments solely concern health and safety output data. Another trend, reported by eight countries, regards the intensifying of sectorial data production. The most interesting of all developments have been pointed out as 'good examples'.

Furthermore, it has been concluded that the lack of comprehensive, reliable (statistical) information at national level is increasingly identified as a major problem. This is particularly with regard to national governments who recognise the limited ability for setting national priorities and guiding interventions in the working environment. Social partners seem not so much concerned by these deficits. Several causes for the deficiences have been mentioned by interviewed key informants, who also identified small and medium sized enterprises as a group in need of special attention, also in data gathering.

Policy options for improvements

Considering some European figures, such as 8000 reported fatal accidents, 4.5 million reported accidents with work interruption, and an estimated 20,000 million ECU paid each year to compensate for occupational injuries and illnesses, as well as considering the observed deficits in the data, such as the significant under reporting and lack of cross-national comparability, it has been concluded that further improvement actions are desirable. Actions both aimed at improving the working conditions, and at improving the monitoring thereof.

Recommendations on the working conditions across Europe
It is suggested to prioritise preventive and improvement actions which are aimed at the indicated main and common risk factors, and the common risk groups across Europe:

- working postures/movements, overload/extension of the body, musculo-skeletal disorders;
- psychosocial hazards, i.e. job content, working pace, influence and control, other stress-related hazards;
- safety hazards related to slips, trips and falls;
- noise;
- agents causing skin diseases, neoplasms and disorders of the circulatory system;
- new technology related hazards;
- male workers, particularly in relation to noise, vibrations, occupational accidents;
- workers in the (metal)manufacturing industries, building and engineering sectors, in relation to occupational accidents.

Suggested actions at European level are subsequently: literature-based in-depth analysis of the extent and severity of the problems and of best practices applied; confrontation of the analysis' results with current EU-policy and development of new policy if needed; European campaigns consisting of a variety of improvement actions aimed at national, sectorial and company level; campaigns' evaluation. To ensure success it is recommended to involve representatives from governments, employer's organisations, unions, health and safety experts and other experts, both at European and national level. It is also suggested to attune these actions to current developments in national working populations, e.g. ageing of workers and increasing participation of female workers.

Recommendations on occupational health and safety monitoring across Europe
Firstly, it has been stated that the most obvious recommendations in this respect simply flow from the deficitis in the European data observed, when taking the 'theoritical ideal monitoring system' as reference: make available data which proved to be unavailable, improve the quality of the data which are unreliable, and harmonise data which are cross-nationally incomparable. In the study report, the topics concerned have been specified. However, these improvement activities have not been put forward as the ultimate recommendations of the study, because they seem not entirely practical. Instead, as a main recommendation it is suggested to carry on a discussion on what would be the desirable and feasible European monitoring system. In addition to this long term European strategy, recommendations are also given for a short term European strategy, and for national initiatives. These are specifed hereafter.

Long term European strategy
Various subjects have been suggested to serve as ingredients for the discussion on the European monitoring system, such as: the crucial matter of which policy purposes the system should fulfil; the hereto related options of having a maximal, optimal, minimal, or no European monitoring system, or a growing scenario in this respect; how to address small and medium sized enterprises more adequately; the role of social partners in monitoring improvement; balancing what needs to be known in order to set priorities, and the amount of effort this security demands; the creation of win/win situations for national and European data producing institutions.

Furthermore, several concrete activities have been suggested to facilitate the recommended discussion: to carry on a scenario-study to work out the various options for a European monitoring system, and to assess their advantages and disadvantages; assigning a working group of policy makers and experts on data production to carry on the discussion and prepare a strategy proposal on European monitoring; linking a supportive network of institutions involved with data production to this working group, e.g. to work out specific aspects of the strategy.

Short term European strategy
Short term activities could be set to work parallel to the long term activities in order to realise motivating successes at short notice, or could be initiated if European policy makers decide against a long term strategy. Various activities have been suggested, both in relation to the monitoring in general, as well as concerning the three monitoring areas separately.

The first general suggestions regard: to apply a general procedure to set European priorities to working conditions improvements, by carrying out studies (e.g. compiling European profiles of the ten presumed most hazardous sectors) for rough problem identification, followed by in-depth studies on prioritised risks and risk groups to assess 'best interventions', and concluded by assessing how to implement the interventions. Other general suggestions are: to assign existing or new working groups of experts from the main data producing institutions, to improve and attune the methodologies on specific monitoring topics; and to enable further analysis of the material obtained in this study, which was left to future analysis because of time restrictions.

With respect to monitoring the working environment it has been suggested to up-date and improve the Eurosurvey in a specified way, because this seems the best option to get cross-national comparable data at short notice and with reasonable efforts. This could be followed by the two last steps of the general procedure suggested above. Furthermore, it is suggested to actively stimulate and support at European level the production of reliable and comparable data on the identified future risks, i.e. new technoloy and stress.

Interventions are suggested to be 'monitored' more intensely, but rather in a qualitative way than quantitatively. It is suggested to establish a European database (paper or electronic) of practical success stories on innovative, inspiring policy interventions to prevent, reduce and control occupational health and safety problems. This database could serve as a source of inspiration for policy makers at European, national, sectorial and company level.

In relation to monitoring health and safety output a suggestion is to consider harmonisation initiatives on specified output topics, after more in-depth having analysed the various cross-national comparative studies on these topics from the recent past. Particularly if such harmonisation initiatives are not feasible, it is suggested to consider the setting up of a cross-nationally standardised trailer questionnaire to the national Labour Force Surveys, covering health and safety output topics. Development and testing of a standardised method to estimate all economic costs involved in occupational health and safety in the various European countries is another recommendation.

National initiatives
The final set of recommendations regard actions which national policy makers and data producing institutions could undertake, attuned, if and where applicable, with the actions at European level. Much of these suggestions mirror those at European level: discuss the desirable and feasible national monitoring system in relation to its desired policy purposes; identify the topics on which data should be made more available, reliable and cross-nationally comparable (some specific suggestions have been given to a number of countries); participate in international platforms, such as networks and working groups, for exchange of information and joint activities on monitoring methodologies; stimulate the relevant European organisations to financially and otherwise support (cross-)national initiatives to improve occupational health and safety monitoring across Europe.

1 Introduction

1.1 Context of the project

Initiatives on health and safety at the work place are steadily moving from the national level to the European level. In order to identify where preventive action on occupational health and safety is needed, it is important to develop comparable and homogeneous data and research on this subject at European level. The need to intensify the monitoring of health and safety in the European Union (EU) was clearly signalled in the European Council's resolution of 21 December 1987 on health and safety at work. The creation of the internal market and the implementation of the social dimension has further increased this need (European Foundation for the Improvement of Living and Working Conditions, 1994a).

The European Foundation for the Improvement of Living and Working Conditions (the Foundation) took immediate action on the Council's request for improved monitoring of occupational health and safety in Europe. From 1988 on, the Foundation has gradually intensified and extended its work on monitoring health and safety at work in the EU. Much of this work has been carried out in close cooperation with the social partners, government authorities, the European Commission, and health and safety researchers and practioners (European Foundation for the Improvement of Living and Working Conditions, 1994a). A brief overview of the activities, carried out by the Foundation since 1988, in order to improve (the monitoring of) occupational health and safety, is listed below. A more elaborate overview can be found in Euro Review, the Foundation's half yearly bulletin on research in health and safety at work (see under c.).

a. Foundation's activities regarding monitoring of working conditions:

- An information booklet was published, presenting an overview on how occupational accidents and diseases are reported in the EU (European Foundation for the Improvement of Living and Working Conditions, 1988).
- The European Safety and Health Database 'HASTE 'has been developed during the period 1988-1993. It has been used by the Foundation to identify monitoring methodologies common in the various EU Member States. This database contains descriptions of over 160 information and monitoring systems on occupational health and safety in EU Member States and in international organisations (European Foundation for the Improvement of Living and Working Conditions, 1995a).
- Two European Conferences on Monitoring the Work Environment were held to discuss the state of the art, and the design of the next steps towards unification of monitoring systems in the EU. The first conference was held in November 1990, the second in November 1992. Policy makers, as well health and safety practioners from the EU, Nordic countries, Eastern Europe, USA, WHO, and ILO were among the participants. (European Foundation for the Improvement of Living and Working Conditions, 1994a).
- A first European questionnaire-based survey on the work environment was held in 1991 among a representative sample of workers in all EU Member States (European Foundation for the Improvement of Living and Working Conditions, 1992). A second one is scheduled for 1996.
- A number of European networks and working groups have been established. These networks and working groups analyse the conditions for exchange of information and for harmonization of the monitoring of health and safety in Europe. Networks have been working on questionnaire-based surveys (Netherlands Institute for Preventive Health Care - TNO, 1994), and on product and exposure registers (European Foundation for the Improvement of Living and Working Conditions, 1994c).

A working group has been operating on workplace assessment.

- As an experiment, a project with a sectorial approach to monitoring and improving working conditions was carried out in 1993-1995. This study, called 'Monitoring the Work Environment at Sectorial Level' resulted in an overview of the health and safety situation in the Meat Processing Industry, and in the Hospital Sector across ten EU Member States (European Foundation for the Improvement of Living and Working Conditions, 1995b and 1995c).

b. Foundation's activities regarding instruments for preventive action:

- A project on sickness absenteeism in the EU was carried out in 1993-1995, called 'Ill-health and Workplace Absenteeism'. Besides providing quantitative information on absenteeism and disability in the EU Member States, the project's aim was to summarize the main characteristics of the national legislation concerned, as well as to describe some good examples of actions taken at company level to reduce absenteeism (European Foundation for the Improvement of Living and Working Conditions, 1994d).
- A research on the 'Identification and Assessement of Occupational Health and Safety Strategies in Europe' was held in 1994-1995, which mainly assembled both quantitative and qualitative data on various policy instruments used in EU Member States to support and stimulate improvement of working conditions (European Foundation for the Improvement of Living and Working Conditions, 1995d).
- A study on economic incentive models was carried out in 1993-1994 with the aim of stimulating enterprises to improve the quality of their working environment. A catalogue of models was compiled which provides elements for an ideal European economic motivation model (European Foundation for the Improvement of Living and Working Conditions, 1994e).
- A working group has been working on monitoring stress at the workplace, and on assessing the benefits of stress prevention.
- A handbook is being developed regarding designing for integration of people at the workplaces, comprising ergonomic design of work stations and buildings, work organisation and processes, time patterns, training etcetera.
- Also a prototype for an improved computerised design guide is under development, along with the identification of information sources and methodologies for healthy design.

c. Foundation's activities regarding dissemination of information:

- A prototype information bulletin on occupational health and safety research in Europe is published and disseminated half yearly: the previously mentioned Euro Review with its Issues 1994-1 and 1995-1 (European Foundation for the Improvement of Living and Working Conditions, 1994b). As well as providing an overview of the trends in occupational research, each issue contains in-depth information on specific topics, such as repetitive strain injuries, organic solvents and work related allergies.

Parallel to these activities of the Foundation, the European Commission has been and still is working intensely on the question of harmonization of occupational health and safety statistics. For example by enacting its Recommendations to the Member States as early as 1962 and 1966, and more recently in 1990, concerning the adoption of a European schedule of occupational diseases (European Commission, 1962, 1966, 1990). The Commision has also been active in providing insight in the availability of occupational exposure data in the European Community (European Commission, 1993).

Furthermore, the Commission has launched two projects which are being carried out by Eurostat, on the collection of comparable data concerning accidents at work and occupational diseases, based on national registers usually related to private or public insurance systems in most of the EU Member States ('ESAW-project' and 'EODS-project'; Eurostat 1994a, 1994b). The aim is to furnish authorities with information for the establishment of national priorities for the prevention of occupational accidents and diseases. Because problems regarding comparability of data were foreseen, the projects will have to map out differences between countries in the operational methodologies, particularly under-reporting, and coverage of groups. For the occupational diseases, the diagnostic criteria for differentiating between work-related and non work-related diseases will also be mapped out.

In relation to the ESAW-project, the first comparable data on accidents at work was transmitted to Eurostat at the beginning of 1995. The data now covers the most basic variables, which are scheduled to be extended by the year 2000. With respect to the EODS-project, the Commission has recently published an information notice on the recognition of the most widespread occupational diseases (European Commission, 1994). Furthermore, a pilot project on cases of occupational diseases recognised in 1995 is now being carried out, which will result in data being available in 1997. The number of variables and diseases, now covering only a subset of items from the European Schedule of Occupational Diseases, are scheduled to provide full coverage by the year 2000.

In addition to the Foundation and the Commission, governmental and expert organizations in various European countries are also known to have taken initiatives for cross-national research. Initiatives in the Netherlands include for example, a comparative research on sickness absenteeism in the Netherlands, Belgium and Luxembourg (Prins, 1989), a study on arrangements and data in six European countries with respect to work incapacity (Prins et al, 1992), and a study on sickness and invalidity arrangements with facts and figures from six European countries (Einerhand et al, 1995).

However, interest in cross-national comparable data on occupational accidents and diseases exists not only at European or national level but also on a larger international scale. An example is the International Labour Office (ILO), which yearly publishes the Year Book of Labour Statistics, containing a.o. statistics on occupational injuries. Recently the ILO elaborated a Code of Practice on recording and notification of occupational accidents and diseases. This was discussed by a group of experts in October 1994, and the amended version has been approved by the Governing Body of the ILO (ILO, 1994a). Another example of interest in occupational health and safety at a large international scale, concerns the Organisation for Economic Co-operation and Development (OECD). In 1989 and 1990 it published Employment Outlook reports, which contained qualititative and quantitative data on occupational accidents, illness and diseases (OECD, 1989 and 1990). Other internationally active organisations are the International Agency for Research on Cancer (IARC) and the World Health Organisation (WHO). An example of their activity towards occupational health and safety data is the Atlas on Cancer Mortality in the European Community (IARC, 1992).

A general observation from the above mentioned activities, is that although a number of information sources exist, very little comparable quantitative occupational health and safety data is available at European level, at this stage. It was against this background that, early 1994, the Foundation took the initiative, as a part of their four year programme 1993-1996, to launch a pilot project called 'the European Working Environment in Figures'.

The Foundation considered it useful to try to support the 'soft data' from the First European Survey on the Working Environment 1991-1992 (European Foundation for the Improvement of Living and Working Conditions, 1992) with 'soft and hard data' available from monitoring systems in the (future) EU Member States, from the various Foundation networks and working groups, and from the other projects and publications mentioned earlier. Hence, this pilot project is closely linked with the previously mentioned activities of the Foundation, and with the other mentioned activities concerning occupational health and safety data gathering and harmonization.

This consolidated report is the result of the project 'the European Working Environment in Figures' in which information on sixteen European countries has been assembled. The report describes the aims and methodology of the project, the problems encountered, the results, discussion and conclusions on the situation in Europe with respect to figures on occupational health and safety issues. It also contains recommendations for future activities to improve both the working conditions and the monitoring in Europe.

1.2 Aims of the project

In the first place, this experimental study has been carried out to provide a quantitative overview of the working conditions in sixteen countries within Europe. Secondly, the aim was to develop an overview of possible gaps in the existing information on working conditions, by mapping out the availability, reliability and comparability of quantitative data on health and safety at the work place. Furthermore, insight in the current developments regarding 'data production' in this field, was desired, both at national level and at European level. Hence, the specific aims of the study were formulated as follows.

1. To describe the working conditions in Europe based on existing data and information sources.
2. To identify common exposures and risks among workers across Europe, especially in relation to gender, age, occupation and sector of economic activity.
3. To supplement the 'subjective data' from the First European Survey on the Working Environment 1991-1992 with 'objective' data from sixteen European countries.
4. To assess the extent of the lack of comparable data in Europe, and discuss the problems related to this.
5. To identify where further data gathering, quality improvement, or harmonization is most urgent.

This consolidated report summarizes the gathered information, and tries to meet the objectives mentioned above, in order to comply with its main purpose: to provide a tool for policy makers, primarily at European level, but also at national level. It may enable them to discuss the situation with respect to occupational health and safety and the monitoring thereof in their country, and/or across Europe more fruitfully. Such discussions may further stimulate the development and implementation of future policies regarding data production, and/or policies on the improvement of the working conditions itself.

The report may however be of interest to a broader audience, e.g. anyone else interested in figures on the working environment and its output in various European countries, such as statisticians, scientists, researchers and other experts in occupational health and safety.

Furthermore, this pilot project and its report may proof to be a starting point for similar studies in the future, after an evaluation of the experiences in the project and its results. Such periodical updates may then, together with other research activities and materials, provide a substantial contribution to the monitoring of the working environment in Europe at a regular basis.

1.3 Structure of the report and remarks to the reader

Chapter 1 describes the background and context of this study, as well as its objectives. It also presents the outline of the report, thus providing a brief description of the contents of the report.

Chapter 2 will elaborate the methodology of the study. Hence, it will describe how the work has been carried out, both with respect to the national data gathering, as well as the consolidation of the material obtained. It will point out particularly which issues have been focussed on, how they were structured, and which main problems have been encountered. It also describes briefly the positive aspects, as well as the deficits of this research and its results.

Chapter 3 provides an introductory overview of the labour market in the sixteen countries that were studied, thus providing background information on the population concerned in this study. Some main characteristics will be described, by presenting quantitative data on the European working population, as well as the working populations in the various countries. The data will refer to the national working populations as a whole, but they are also presented according to gender, age, occupation, sector of economic activitity, and company size. Furthermore, figures on companies are given.

Chapter 4 is one of the central parts in the report. It gives an overview of the availability, reliability and comparability of data on the working environment and on health and safety output, from national information sources in the countries studied. This is not only done for data regarding the entire national working populations, but also for data according to gender, age, occupation and economic sector. Furthermore, it provides an overview of the information sources from which these data stem, and the organisations involved in data production in the various countries.

Another main part of the report is chapter 5. It presents both figures, as well as qualitative interpretations on topics on which data are best available across Europe, i.e. available in six countries or more. With respect to the working environment, these topics concern physical exposure (noise, vibrations, climate, radiation), exposure to psychosocial risk factors (working pace, job content, working hours, influence and control, social interaction), and exposure to physiological hazards (working postures and movements, manual handling). In relation to the health and safety output, information on five topics is presented: occupational accidents, sickness absenteeism in general, occupational diseases, occupational mortality and general mortality. Where sufficient data was available (from six countries or more), and as far as is methodologically justifiable, not only information regarding the total national working populations is given, but also broken down to age, gender, occupation and economic sector. From this information, preliminary conclusions are drawn with respect to common and main risk factors, main risk groups, and main problems with respect to health and safety output across Europe.

Current trends and strategies regarding occupational health and safety data production are the subjects of chapter 6. Here, topics will be pointed out which have been identified as main problems by the government, social partners and other relevant organizations in the various countries. The topics are described which have been chosen by these parties on which to take initiatives. An overview of developments and activities taking place at national level with respect to data production, is also given. The discussion in this chapter will position these results into a European perspective, by referring to the context of the study, as described in section 1.1.

Chapter 7 is a synthesis of the discussions and preliminary conclusions of the previous chapters 4 to 6. It consists of an overall discussion and final conclusions. These review the availability and the quality of data on the one hand, and the working environment and its health and safety consequences, on the other hand. Hence, common risk factors and main occupational health problems across Europe are identified, as well as major information gaps and main current developments in data production.

The identification of deficits in the working conditions, and of 'white spots' in existing information will lead to recommendations for future actions. These are taken up in chapter 8. Here, priorities for improvement of the working environment are presented, as well as priorities for further data gathering, quality improvement, and harmonization. These recommendations can be considered to be policy options, both at European and at national level, with respect to future monitoring of occupational health and safety across Europe.

Chapter 9 consists of the bibliography of information sources mentioned in this consolidated report. It does not contain the large number of information sources of each national report which were used to compose this consolidated report. Details of these national sources can be found in the national reports themselves. Annex 1 provides a list of organisations and contact persons from whom these national reports may be obtained.

The Annexes conclude this report. They particularly provide explanatory and background information on the methodology of the study.

A last remark concerns readers who can not find enough time to read the entire report. To them it is suggested to restrict themselves to the Summary, chapter 7 'Final conclusions and general discussion' and chapter 8 'Recommendations for future policy'. There they will find all the headline information this report contains.

2 Methodology of the study

2.1 Theoretical monitoring model

As a starting point for the study, a theoretical model for occupational health and safety monitoring was discussed within the Steering Group, which was established for the purpose of this project. Including all the researchers, it also consisted of one representative from the Foundation, two from national governmental bodies, two from social partners, one from the European Commission/Eurostat, and one from one of the research institutes involved (see Acknowledgements).

This theoretical model is taken up in figure 2.1 and described thereafter.

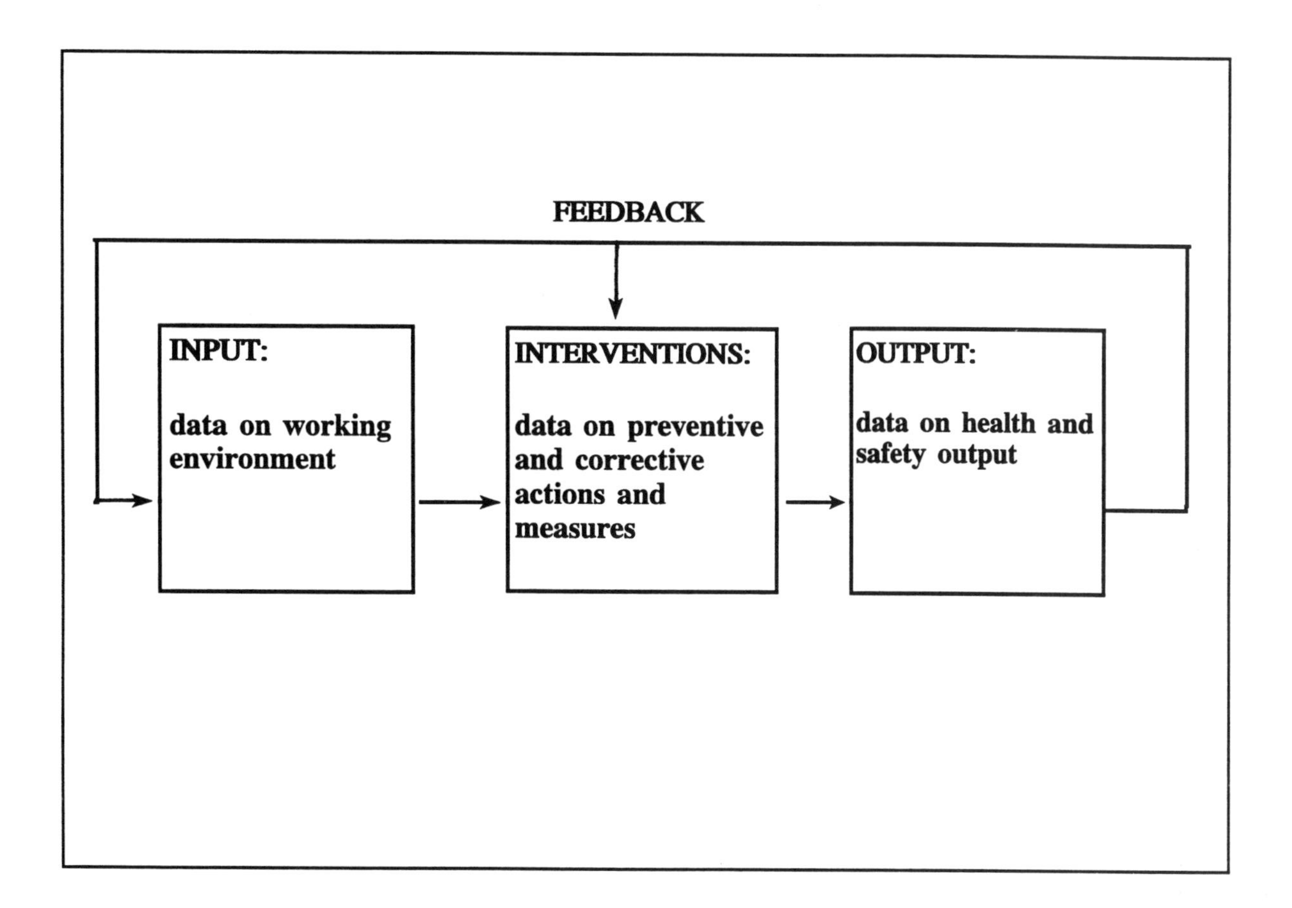

Figure 2.1 ***Structure of a 'theoretically ideal' monitoring system to monitor occupational health and safety at European level***

Theoretically spoken, a European monitoring system ideally should cover three areas concerning occupational health and safety:

a. Input: data on exposure to all risk factors in the working environment, such as physical factors like noise, psychosocial hazards like lack of influence and control over own work, or biological agents like viruses and bacteria.

b. Interventions: data on all actions and measures undertaken to reduce the exposure to risk factors and to improve the working environment. This concerns for example data on inspection and enforcement activities, data on the proportion of the working population covered by Occupational Health Services, and data on actual work place improvement actions at company or sectorial level.

c. Output: data on all kinds of health and safety consequences, such as occupational accidents, diseases, and sickness absence, as well as data on economic costs involved.

Ideally, sufficient data should be available within each area, not only on the entire working population involved, but also broken down to gender, age, occupation, economic sector or any other category considered relevant, to enable identification of risk groups or target groups for action. This would mean that a substantial number of European countries should have data available to fill the monitoring system. In this study, this minimum requirement was more or less pragmatically set on six out of sixteen countries.

Ideally, the available data should further be reliable (i.e. valid, reproducable, representative, objective or 'hard'), up-to-date and cross-nationally comparable. Furthermore, the data should inter-relate in order to be able to link causes, interventions and effects.

Figure 2.1 further indicates that the data flow from each area finally results in feedback of information which may result in (hopefully positive) changes in the situation of each area: exposure levels or number of exposed people may be reduced, interventions may be intensified or altered in nature, and the number or severity of e.g. occupational accidents may decline. This feedback is rather crucial considering a general purpose of a monitoring system: to provide steering information for policy making on safe and healthy work for people.

2.2 National data gathering

The study was set up to assemble information from sixteen European countries: the fifteen EU Member States and Norway. Hence, the following countries are included in this research (country codes used in the text hereafter are put between parentheses).

Austria (A)
Belgium (B)
Denmark (DK)
France (F)
Finland (FIN)
Germany (GER)
Greece (GR)
Italy (I)
Ireland (IRL)
Luxembourg (L)
Norway (N)
The Netherlands (NL)
Portugal (P)
Spain (E)
Sweden (S)
United Kingdom (UK)

The data gathering for each country was delegated to five research institutes from five different countries, according to the following clustering.

1. Belgium, Austria, France, Germany and Luxembourg.
2. Denmark, Finland, Norway, Sweden.
3. Spain, Greece, Italy, Portugal.
4. United Kingdom, Ireland.
5. The Netherlands.

In order to obtain the necessary information, researchers from these five institutes contacted institutes and persons in the countries they had to screen. Therefore, most data are in fact obtained from national institutes in the sixteen countries. In addition, some data were obtained from Eurostat (see below). Annex 1 provides an overview of the research institutes and contact persons whom participated in this study.

For the purpose of the study, a so-called 'matrix' was developed, along with explanatory guidelines, which was based on the theoretical monitoring model. This development, which took place from March till June 1994, was carried out and co-ordinated by researchers from the Netherlands. This work was based on contributions from all researchers plus Eurostat. Support to this development was further given by discussions within the Steering Group.

The matrix provided a uniform structure for the data gathering and reporting on each country. The structure consisted of the following sections (see Annex 2 for the full contents of the matrix, including all the items studied).

1. Description of the context.
2. Description of the working environment.
3. Health and safety output.
4. Trends and strategies regarding data production.

While developing the matrix, it was discussed whether data on preventive and corrective actions (interventions) and on economic costs should be gathered as well. It was decided not to include the topics in the matrix, because there would be overlap in the preventative actions with the parallel running project 'Identification and Assessment of Occupational Health and Safety Strategies in Europe'. Hence, such data would be available in the consolidated report of that project (European Foundation for the Improvement of Living and Working Conditions, 1995d). In relation to the economic costs, the subject was judged to be too complex to incorporate in this study, although it was considered highly relevant to policy makers at European and national level.

Therefore, the main scope of the matrix, being the health and safety monitoring instrument used in this study, concerned the working conditions, the health and safety output, and the developments regarding data production. The various items taken up within these main sections were selected on basis of consensus within the Steering Group: the items were considered to be the relevant and generally accepted items within each section (see Annex 2).

In order to fullfil the second aim of the study (see section 1.2), the matrix not only 'asked for' figures on the entire national working population, but also required information according to gender, age, occupation and sector of economic activity. To increase the chance of obtaining reliable and comparable information, mostly internationally accepted classification systems were used in designing the matrix. They include the following topics and systems:

- occupation categories: ISCO-88 (COM), two-digit level (see Annex 3);
- economic sectors: NACE-1970, one-digit level (see Annex 4);
- risk factor categories: Exposure Classification System - a Proposal (Laursen et al, 1994) (see Annex 5);
- diagnostic categories for diseases: almost entirely corresponding to ICD-version 10, three-digit level (see Annex 6).

The aim of the study was to strive for predominantly quantitative, most recent data, preferably from 1992 or 1993. The matrix however allowed some flexibility in this respect. With respect to the reference period of data, it was allowed to use data from earlier years, if that was the only data available. Reference periods from which the data stems were requested in order to be able to correctly compare data between countries in the consolidation.

Flexibility regarding non-quantitive data was provided by leaving room for 'Remarks/discussion' under the pre-structured tables in the matrix. Here, relevant qualitative information could be taken up, either to supplement, specify, or explain the figures in the table, or to give any other explanatory information, for example on unavailability or reliability of data. More specifications on the information at 'Remarks/discussion' that was requested, can be found in Annex 2.

As a general procedure, the national data had to be gathered from already existing national or international information sources, such as statistics, registers, surveys, other authoritative sources, and professional literature (see Annex 2). In order to trace the various information sources to be consulted, researchers could utilize the Foundation's HASTE Database, containing over 160 descriptions of information systems in EU Member States (European Foundation for the Improvement of Living and Working Conditions, 1995a). They could also use the 'Review of Surveys' (European Foundation for the Improvement of Living and Working Conditions, 1994f), which was especially carried out as a preparation for this study. It forms a catalogue of general, regional and sectoral surveys in the twelve EU Member States, Norway, Sweden, Finland, Austria, some other countries, and from international organisations.

As was mentioned before, the national data were mostly obtained through national institutes, which traced and selected the information sources. There were however two exceptions. Firstly, almost all data of section 1 in the matrix were provided by Eurostat, for the twelve countries being EU Member States at the end of 1994. This information was provided through a print-out, which was especially run for this study, and which was based on the Labour Force Survey 1992 (Eurostat, 1992a). Secondly, the data in section 1 of the matrix with respect to companies were in fact provided for all sixteen countries studied. These data were obtained from the preliminary version of 'Enterprises in Europe, third report' (Eurostat, 1994c), which was also provided by Eurostat.

An exception with respect to utilizing existing information sources for the national data gathering, concerns the consultation of key informants from governmental bodies, unions, employers' organizations, and other authoritative national organizations, such as centres for statistics. Consultation of these informants was necessary to obtain the data in section 4 of the matrix, particularly if this information was not available as written information in, for example, policy documents. Furthermore, this consultation was recommended in order to get access to information sources, to have the data in the draft version of the national report checked, and to find supplementary information and/or sources, if applicable.

The national data gathering took place from June 1994 till June 1995. Eurostat provided its information between July and October 1994.

The research institutes encountered various problems in the national data gathering. For some institutes it meant great effort to trace the correct national institutes and contact persons, which would be the key entrance to the national information sources to be studied. Apart from being time consuming activities, it mainly caused delays in the delivery of the national reports. Another problem concerned the lack of national information sources in some countries, so that only some international information sources could be consulted.

The main problem however, encountered with respect to all countries to a more or lesser extent, regard the inability to provide reliable data fully according to the matrix. In the first place, in each country data were not available on various issues, or not exactly according to the specified structure, definitions and categories of the matrix. This was either due to non-existence of these data, but in other cases it was due to the time restrictions of the project, which for example made it impossible to convert data into the desirable categories. In the second place, reliability of data was often questionned, for example due to known under-reporting, to 'subjectivity' of data, or to the size or nature of samples, which did not allow extrapolation of data to the entire national working population.

Despite these obstacles, national reports from all 16 countries were obtained though. However, the report on Norway only provided information on the working environment and no health and safety output data, whereas the report on Portugal was rather lean regarding the output data (only data on occupational accidents). Nevertheless, all national reports were judged to be of sufficient quality, so that they have all been used in the consolidation.

2.3 Consolidation

As well as providing support to the national data gathering and reporting, the uniformity of the matrix was chosen in order to facilitate the synthesis and comparison of the national data, and thus to facilitate the consolidation of the sixteen national reports into this overall report. The consolidation mainly took place from April till November 1995, after having discussed the outlines of the consolidated report within the Steering Group in February 1995.

To ensure that data from the national reports were correctly presented and interpreted, two draft versions of the consolidated report were sent to all researchers for comments. The other members of the Steering Group received the draft versions as well, for a more general critical check. After processing the written amendments from all these persons as well as the comments from an evaluative meeting, the report was finalized.

As might be understood from the description of the problems encountered in the national data gathering, the consolidation consequently was affected by them. Firstly, the unavailability of data has limited the number of items on which figures could be presented at European level. Secondly, the differences in definitions and classifications used in the national reporting has impeded the cross-national comparability significantly. Added to this, the problem of the questionnable reliability of much of the data, has meant that the figures, if available, could only be interpreted cautiously and had mostly to be taken as indicative. For a more elaborate discussion on the methodological problems involved is being referred to chapter 7.

The main effect of these problems however is, that the emphasis in this consolidated report has resulted in providing an overview on the availability and quality of data, rather than on providing figures and conclusions on the European working environment and its output. Consequently, it means that the objectives 4 and 5 of this study, as described in section 1.2, are much better met, than the objectives 1, 2 and 3. The recommendations for future policy will consequently be more concerned with (the improvement of) monitoring health and safety in Europe, than pinpointing reliably the main risk factors and the main risk groups in the European working environment, which should get priority in prevention and improvement strategies.

Nevertheless, it has largely been possible to carry out the study as was originally set out (see figure 2.2).

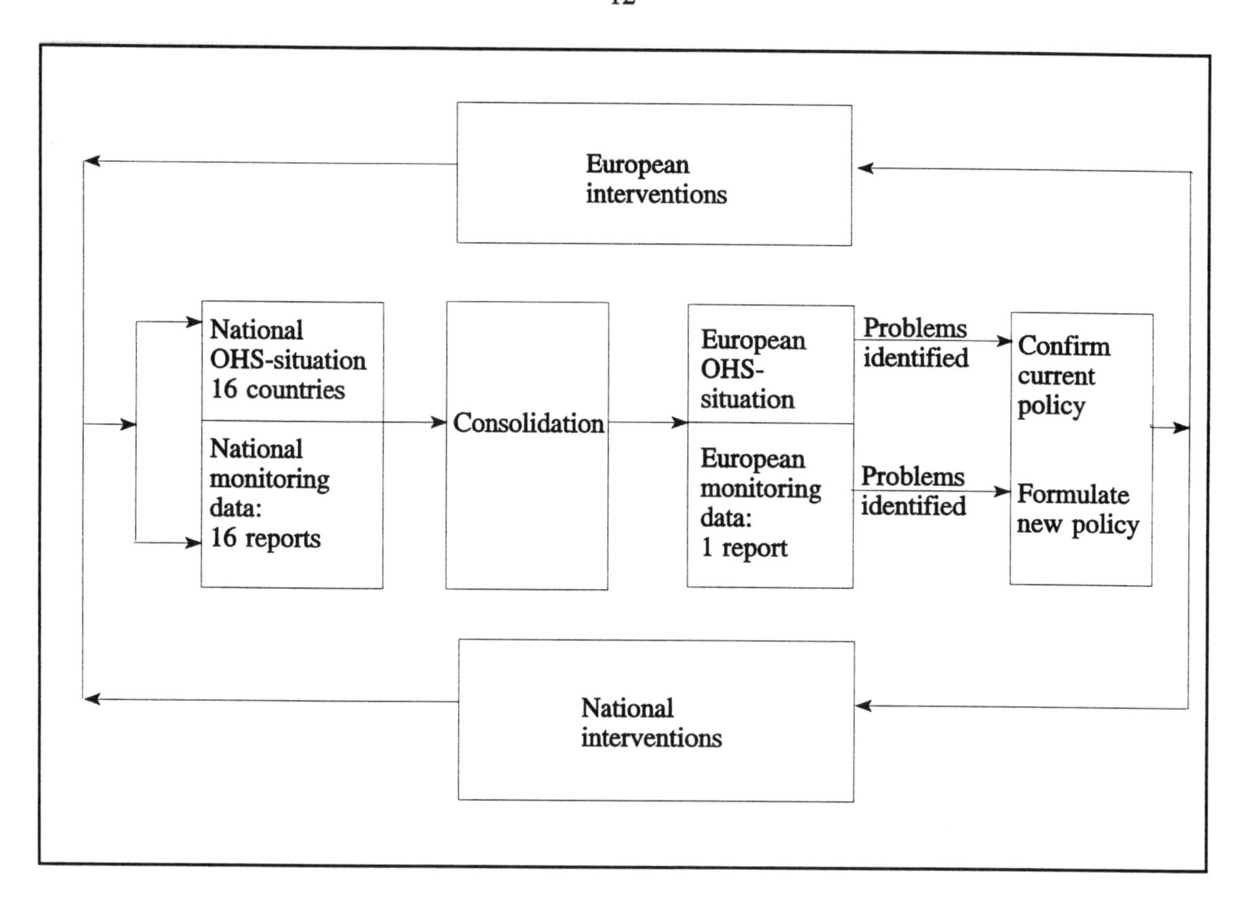

Figure 2.2 *Schematic overview of this study*

As figure 2.2 indicates, the study has resulted in 16 national reports, which not only provide a picture of the occupational health and safety (OHS) situation in these countries, but also give insight in the state of the art of national monitoring. Through the consolidation an impression has been constructed of the same subjects, yet at European level. From this impression, problems have been identified, both in the European occupational health and safety situation, and in the monitoring thereof. Policy makers may confront these problems with current policies in this respect, which can result in either their confirmation, or in feeling needs or desires to formulate new, additional policies. These needs and desires may in turn result in interventions at national or European level, aimed at improving health and safety at work, or improving its monitoring. Initiatives to such interventions were, in the end, the objectives of this study.

Furthermore, this study and consolidated report may, also for other reasons, be considered a valuable step ahead concerning monitoring health and safety at work in Europe. In the first place, because this study, for the first time, provides information from so many countries, and on such a reasonably complete scope of specific items in both the working environment, as well as on health and safety output. Previous studies, as were mentioned in section 1.1, were restricted to a lesser number of countries, and to one or a few items, which were analysed more in-depth, however. In other cases, these studies were more concerned with the monitoring systems, rather than the actual monitoring data.

Significant value of this consolidated report might further come from just this overview of previous studies and initiatives in section 1.1. Although an effort has been made to list as many relevant documents as possible, this overview should however not be considered to present a complete inventory of all available European studies, or current initiatives with respect to occupational health and safety data production in Europe. Moreover, time restrictions in the project made it impossible to thoroughly compare the information from this study, with that from the others, which could have resulted in interesting and useful insights. Yet, this overview of previous studies and initiatives provides quite some points of reference for additional data and analysis for future research.

In this respect, this report presents a new and rather broad synthetical overview of the state of the art of monitoring health and safety in Europe, which can be taken as another stepping stone in European health and safety policy.

3 The European labour market - main characteristics

In this chapter a selection of graphics has been included, describing the main characteristics of the working population in Europe. This selection has been chosen in relation to the matrix which was used to describe various aspects of working conditions. As in the matrix, the volume of the working population per country has been used, including the categories gender, age, occupation and economic sector. In addition a paragraph has been included on the number of companies and percentage of people employed according to company size.

The emphasis of this description is in the comparison of these various characteristics in the European countries. The gender aspect has also been used to produce sub-distributions of the working population according to occupation.

The data are mostly based on material obtained from Eurostat (Eurostat, 1992a; Eurostat 1994c). Data on Austria, Norway, Finland and Sweden are mostly based on their national reports. As far as the distribution by occupational categories is concerned, data on these countries were not available or not obtained in time.

In the graphics the country names are abbreviated as follows:

A	: Austria	GER	: Germany	NL	: The Netherlands
B	: Belgium	GR	: Greece	P	: Portugal
DK	: Denmark	I	: Italy	S	: Sweden
E	: Spain	IRL	: Ireland	UK	: The United Kingdom
F	: France	L	: Luxembourg		
FIN	: Finland	N	: Norway		

Total European working population

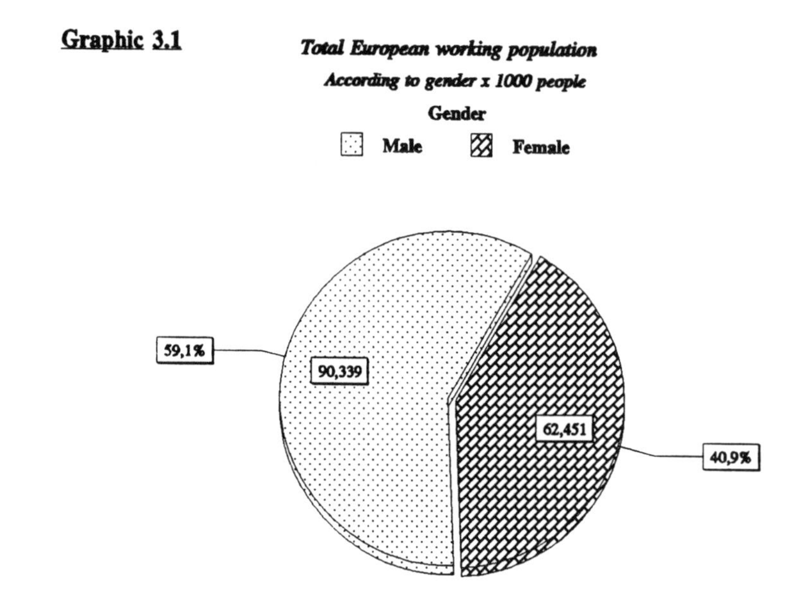

Source: Eurostat, 1992a

Graphic 3.1 gives information about the total European working population according to gender. The total working population of the sixteen European countries is estimated at 152,789,500 people. Graphic 3.1 shows that some 59% of the total European working population consist of male workers and that circa 41% are female workers.

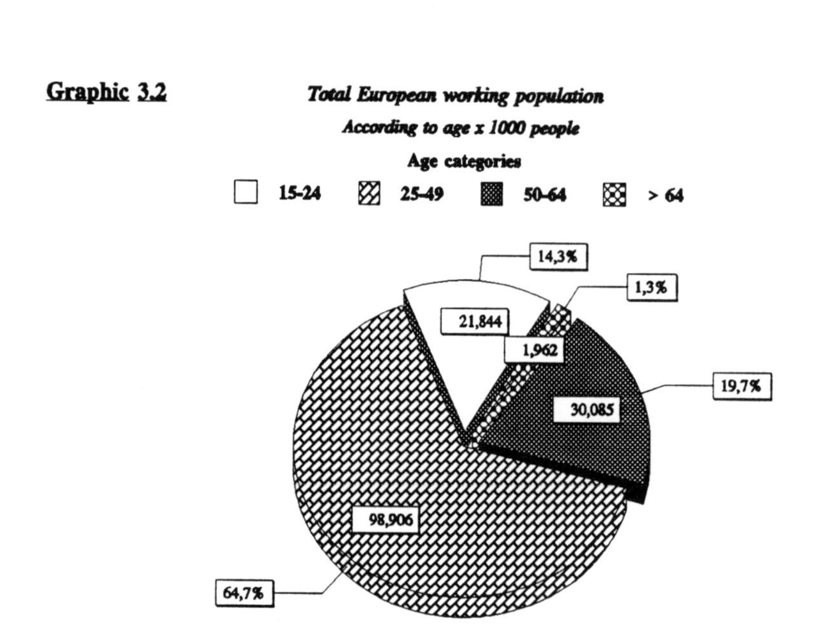

Source: Eurostat, 1992a

Graphic 3.2 shows that the highest percentage of workers are between 25 and 49 years of age (circa 64% of the total European working population). Some 20% of the total European working population are workers between the age of 50 and 64 years; circa 14% are between 15 and 24 years and finally circa 1% of the workers are 65 years or older.

Working population according to age and gender

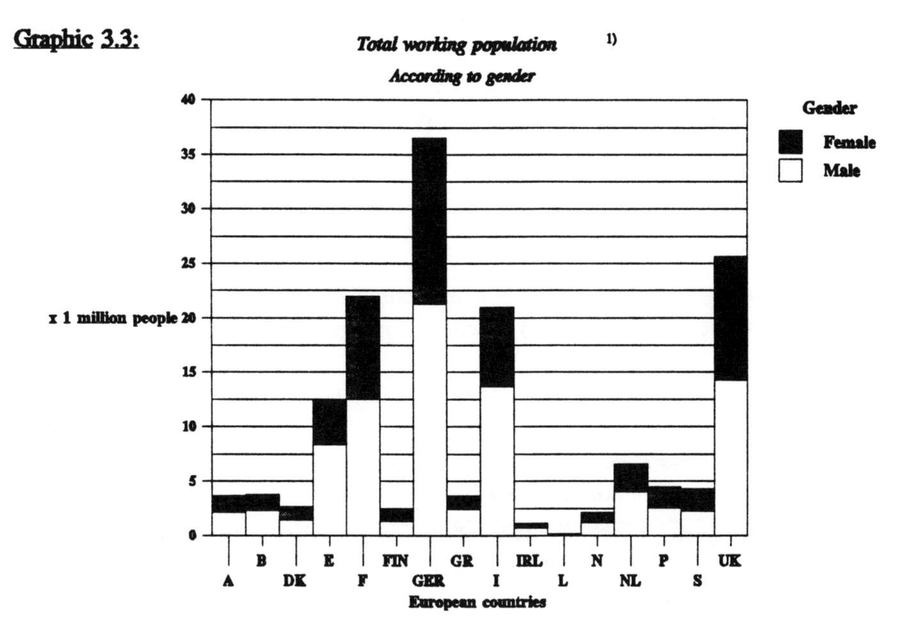

1) Total working population includes employees, self-employed and family workers

Source: Eurostat, 1992a

Graphic 3.3 provides some information on the total working population according to gender for sixteen European countries. It varies between 164,600 (L) and 36,528,300 (FRG). Further research shows that the ratio of the male and female workers does not vary much among the 16 European countries. The percentage of female workers that form part of the total working population varies from 33%(E) to circa 47%(FIN,S).

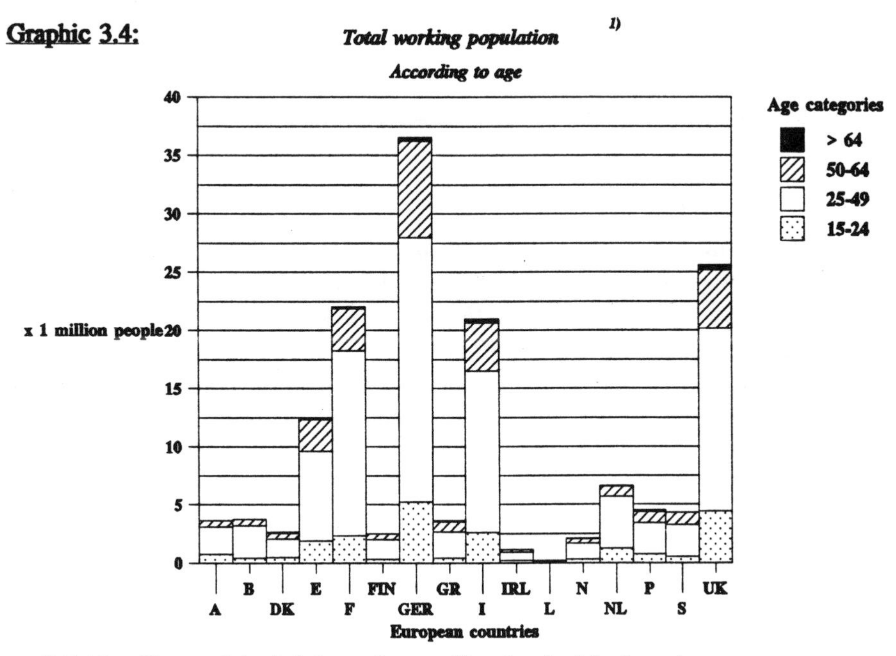

1) Total working population includes employees, self-employed and family workers

Source: Eurostat, 1992a

In graphic 3.4 information is given about the total working population according to age categories. According to age categories the total European working population is divided as followed: 15-24 years (circa 14%); 25-49 years (circa 65%); 50-64 years (circa 20%) and 65 years and older (circa 1%).

Most workers, both male and female, appear to be between 25 and 49 years of age. In most countries this category constitutes 60%(P) to 74%(B) of the total working population. In most countries also the 'older' workers (50-64 years) predominate the younger workers (15-24 years). In five countries (A,IRL,L,NL) it is just the opposite.

Finally, in most countries the age category > 64 years constitutes 0.4%(A) to 2.1%(DK) of the total working population. In Greece, Ireland and Portugal this percentage is higher: 3.1%(IRL) to 3.7%(P).

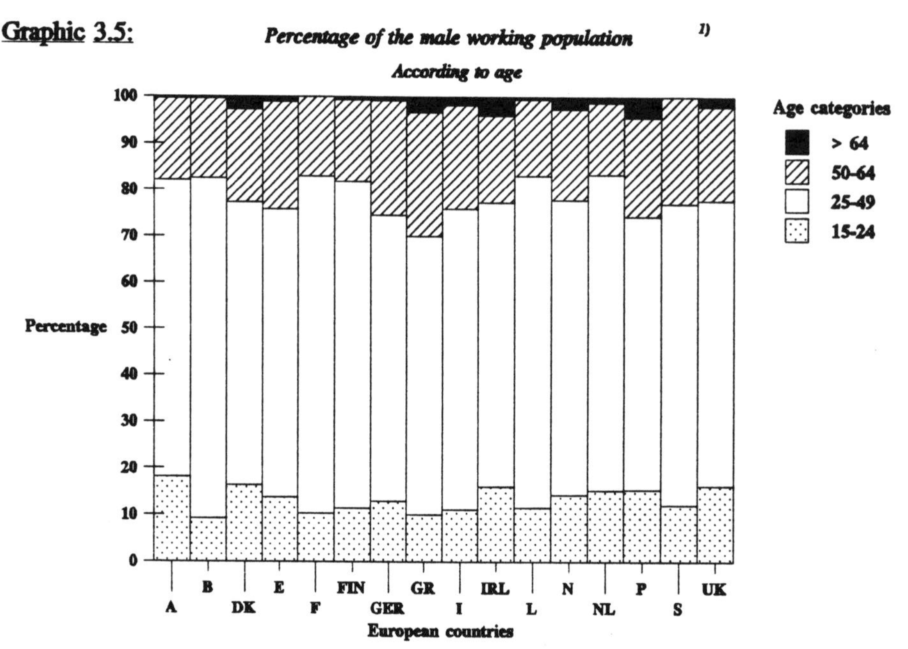

1) The male working population includes all male employees, self-employed and family workers

Source: Eurostat, 1992a

Similar to graphic 3.4 the graphic 3.5 above shows that in all sixteen countries most male workers appear to be between 25 and 49 years of age. The percentage of men (25-49 years) that form part of the male working population varies from 59%(P) to 73%(B). In most countries (except in Austria and the Netherlands) the 'older' workers (50-64 years) predominate the 'younger' workers (15-24 years). Besides that in most countries the category 65 years and older constitutes only 0.4%-2% of the total male working population. This category is larger (3-4%) in Denmark, Greece, Ireland, Norway and Portugal.

Graphic 3.6: *Percentage of the female working population* 1)

According to age

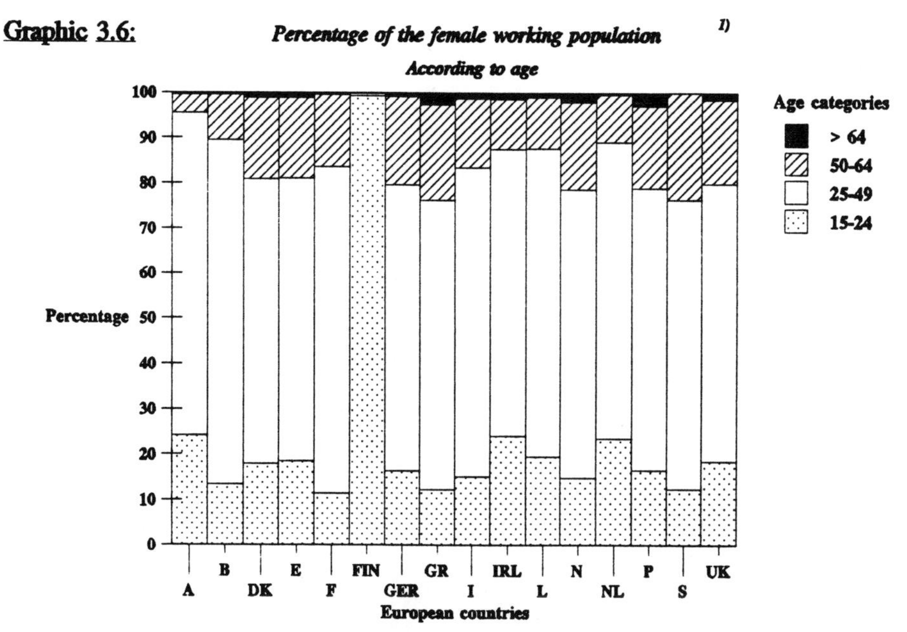

1) The female working population includes all female employees, self employed and family workers

Source: Eurostat, 1992a

Graphic 3.6 shows that in all sixteen countries most female workers appear to be between 25 and 49 years of age. The percentage of women that forms part of the female working population and falls in the age category 25-49 years varies from 61%(UK) to 76%(B).

In eight countries (A,B,E,IRL,L,N,NL,UK) the 'younger' workers (15-24 years) predominate the 'older' workers (50-64 years): it varies from 1%(E,UK) to 18%(A) more younger than older workers. In six countries (F,FIN,GER,GR,P) it is just the opposite (1-11% more older than younger workers). Only in Denmark and Italy the percentages of 'younger' and 'older' workers are the same.

In most countries circa 1% of the total female working population is 65 years of age or older. Only in Greece, Ireland, Norway, Portugal and the United Kingdom this percentage is higher: 2-3% of the total working population of the country.

Working population according to occupation

In respect to the labour force, according to occupation, there is information available from 12 European countries (B,DK,E,F,GER,GR,I,IRL,NL,P,UK). Data from three other countries (A,N,S) was not available or did not correspond with the categories that were used in the matrix. Data from Finland was received too late to be taken up in the graphics.

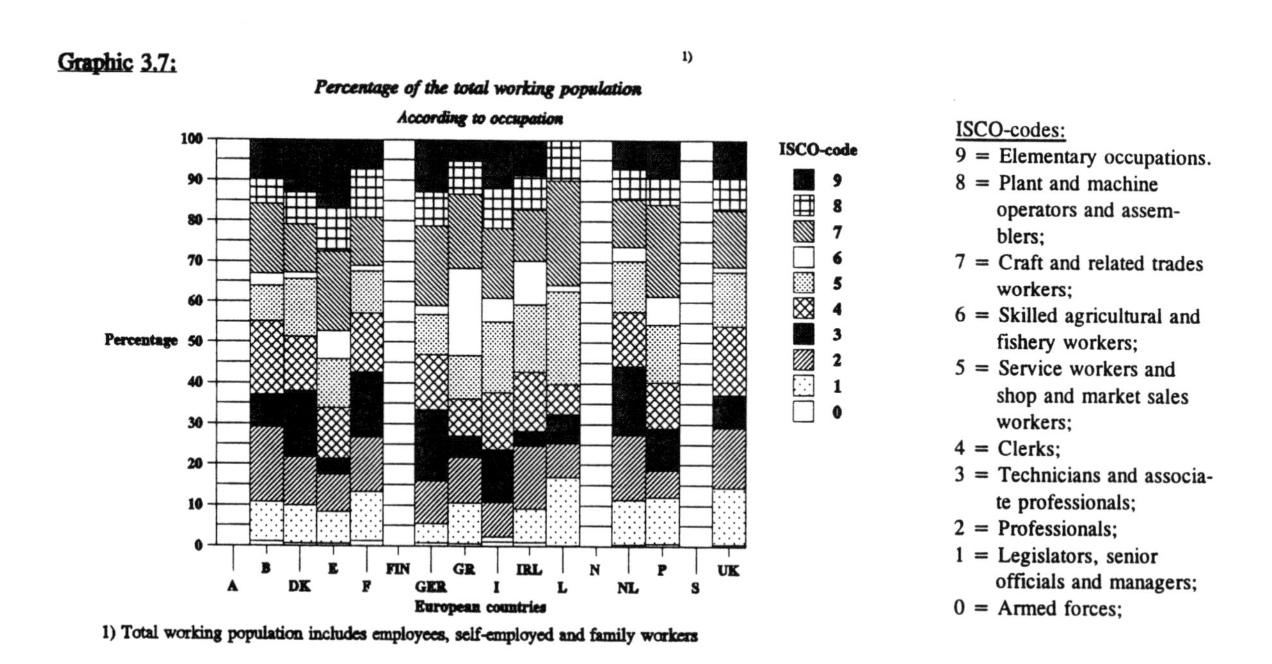

Source: Eurostat, 1992a

In Spain, Germany, Italy, Luxemburg and Portugal most of the workers are employed in the craft and related trades industry: this category constitutes circa 15%(I) to circa 28%(L) of the total working population. In Denmark, France and the Netherlands most of the workers are employed as technicians and associate professionals: circa 16%. Finely in Belgium the professionals (circa 18%), in Greece the skilled agricultural and fishery workers (circa 22%), in Ireland the service workers and shop and market sales workers (circa 17%) and in the United Kingdom the clerks (circa 17%) form the largest profesional group of the total working population.

In all twelve countries the armed forces represent the smallest occupational group of the total working population. This category constitutes only 0-1% of the workers of the total working population.

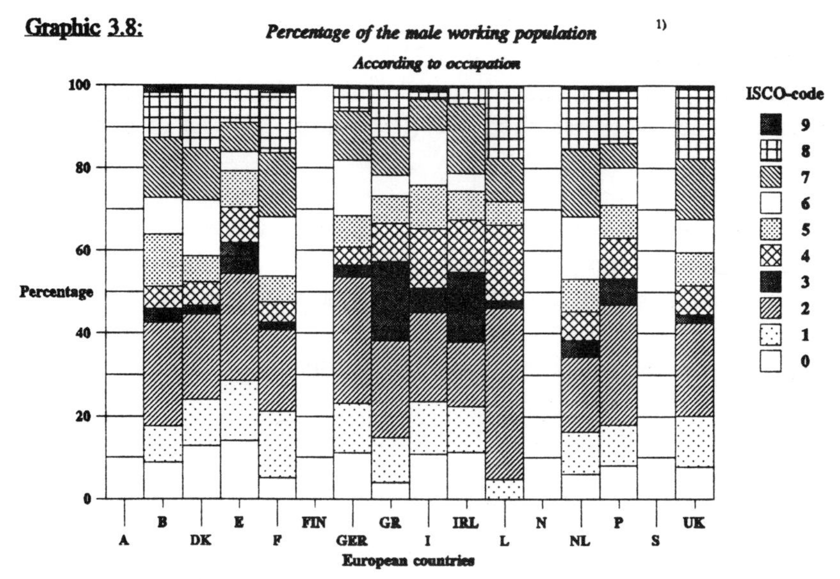

1) The male working population includes all male employees, self-employed and family workers

ISCO-codes:
9 = Elementary occupations.
8 = Plant and machine operators and assemblers;
7 = Craft and related trades workers;
6 = Skilled agricultural and fishery workers;
5 = Service workers and shop and market sales workers;
4 = Clerks;
3 = Technicians and associate professionals;
2 = Professionals;
1 = Legislators, senior officials and managers;
0 = Armed forces;

Source: Eurostat, 1992a

This graphic shows that in eleven countries the male craft and related trades workers represent proportionately the largest professional group. It varies from circa 17%(NL) to circa 41%(L) of the total male working population. In Ireland the skilled agricultural and fishery workers are proportionately the largest occupational group of the male working population (circa 16%).
Likewise in graphic 3.7 the armed forces represent proportionately the smallest occupational group of the male working population. This category constitutes only 0-1% of the male working population. However, in Italy the legislators, senior officials and managers also form 1% of the male working population.

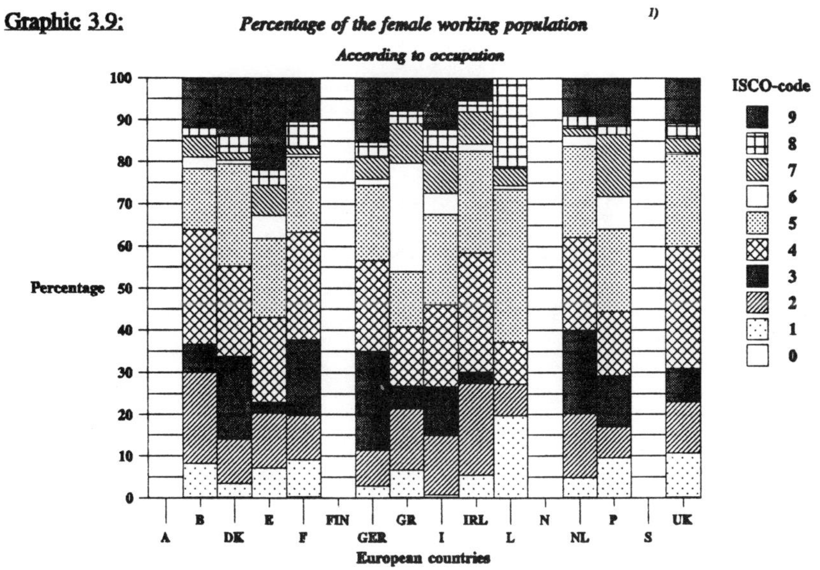

1) The female working population includes all female employees, self-employed and family workers

ISCO-codes:
9 = Elementary occupations.
8 = Plant and machine operators and assemblers;
7 = Craft and related trades workers;
6 = Skilled agricultural and fishery workers;
5 = Service workers and shop and market sales workers;
4 = Clerks;
3 = Technicians and associate professionals;
2 = Professionals;
1 = Legislators, senior officials and managers;
0 = Armed forces;

Source: Eurostat, 1992a

In Belgium, France, Ireland, the Netherlands and the United Kingdom proportionately most of the female workers work as clerks. This category constitutes circa 21%(NL) to circa 28%(UK) of the female working population. In Denmark, Italy, Luxemburg and Portugal this holds for the female service workers and shop and market sales workers: circa 20%(P) to circa 37%(L) of the female working population.

In Spain most female workers have elementary occupations (circa 22%), in Germany most of the female workers work as technicians and associate professionals (circa 23%) and finally in Greece most of the female workers are employed as skilled agricultural and fishery workers (circa 26%).

In all twelve countries the armed forces represent the smallest occupational group of female workers.

Number of companies and percentage of people employed according to company size

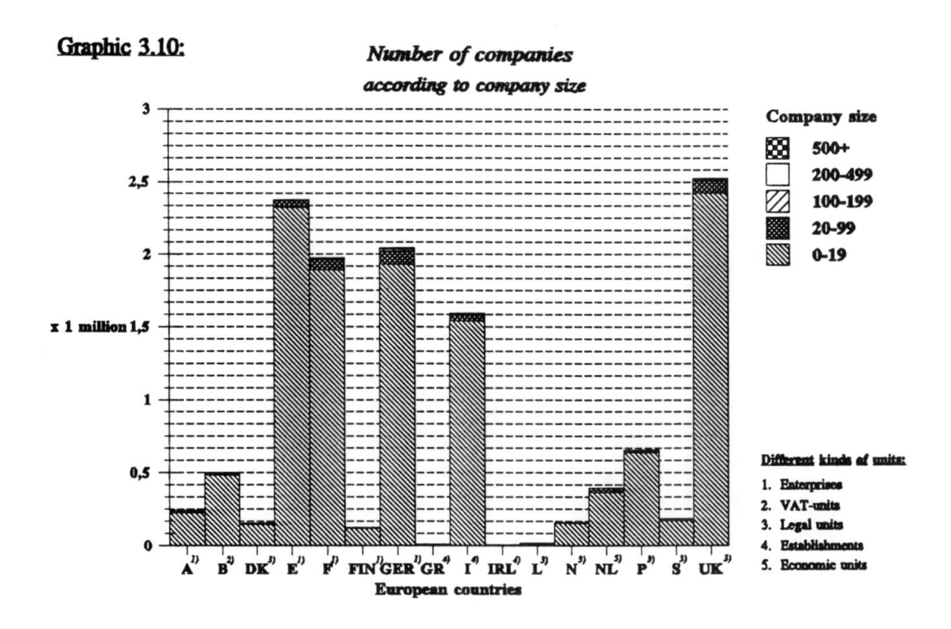

Source: Eurostat, 1994c

Graphic 3.10 shows the number of companies (according to company size) for the sixteen countries. The range of the total number of companies varies from 4,804(IRL) to 2,524,308(UK). Regarding the company size it is clear that in all countries most companies have 0-19 employees. The ratios of the different company size are displayed and further discussed in graphic 3.11.

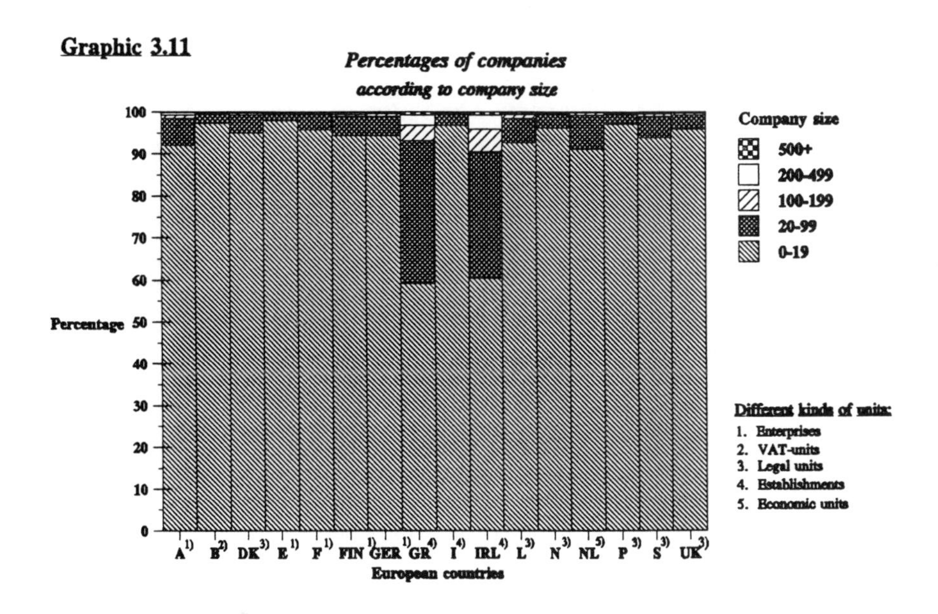

Source: Eurostat, 1994c

Graphic 3.11 shows that in all sixteen countries the most common type of companies is that with 0-19 employees. The range varies from circa 58%(GR,IRL) to 97%(B,E,I,N,P) of all types of companies.

The percentage of companies with 20-99 employees is relatively large in Greece (circa 34% of all types of companies) and Ireland (circa 29%), whereas in the other 14 countries this percentage is significantly smaller: the range varies from circa 2%(B,E,P) to circa 6% (NL).

Graphic 3.12: *Percentage of total working population **

According to company size

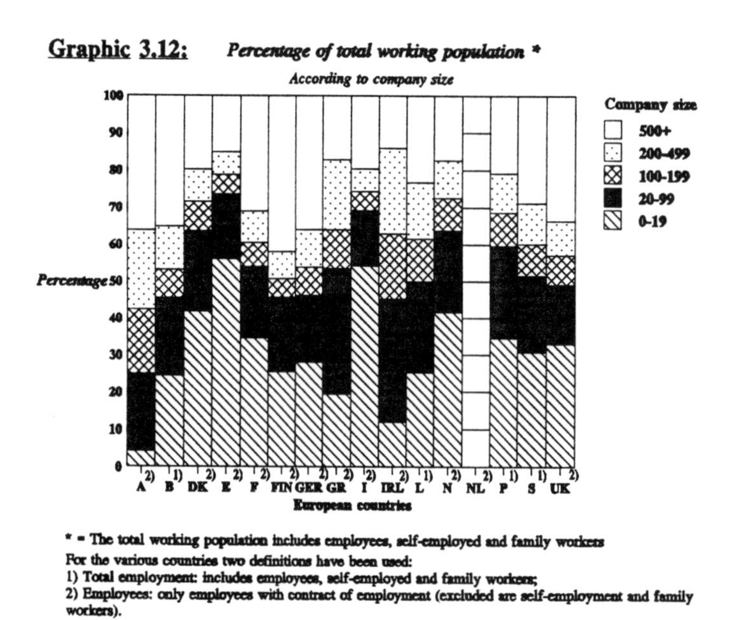

* = The total working population includes employees, self-employed and family workers
For the various countries two definitions have been used:
1) Total employment: includes employees, self-employed and family workers;
2) Employees: only employees with contract of employment (excluded are self-employment and family workers).

Source: Eurostat, 1994c

Graphic 3.12 shows information on the percentage of the total working population according to company size. In eight countries (DK,E,F,I,L,N,P,S) the largest part of the total working population is employed in companies with 0-19 employees. The range of percentages varies from 25%(L) to circa 55%(E,I). Furthermore, in five countries (A,B,FIN,GER,UK) the largest part of the total working population is employed in companies with more than 500 employees. The range of percentage varies from 34%(UK) to 37%(FIN) of the total working population. Finally, in Greece and Ireland the largest part of the total working population is employed in companies with 20-99 employees. In Ireland this counts for 29% of the total working population, whereas in Greece it is 34%. The data provided by Eurostat (1994) on the Netherlands (NL) was incomplete and could therefore not be included in this graphic.

Working population according to economic sector

Graphic 3.13: *Total European working population According to economic sectors*

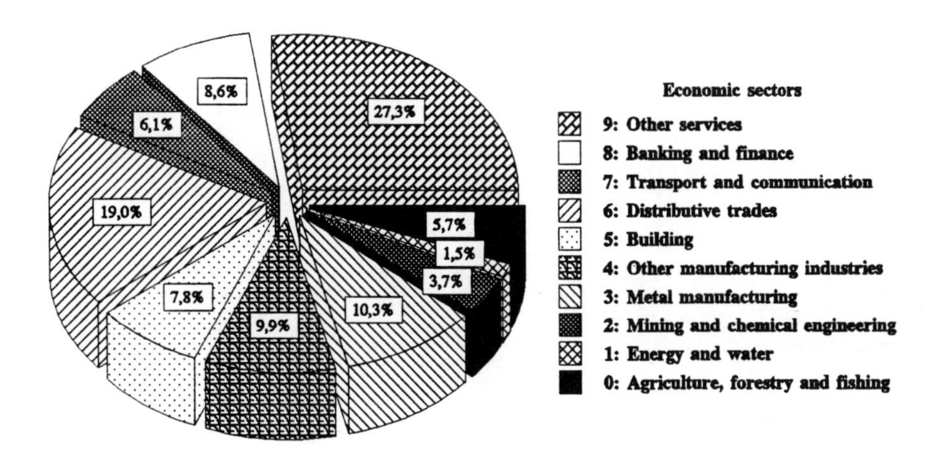

Source: Eurostat, 1992a

Graphic 3.13 shows that in 1992 more than 27% of the total European working population (circa 39 million workers) are working in the sector *'other services'*. The sector *'distributive trades'* forms part of 19% (circa 26.3 million workers) of the total European working population and is therefore the second largest economic sector. Finally, the sector *'energy and water'* is, with 1.5% of the total European working population (circa 2.1 million workers), the smallest economic sector in 1992.

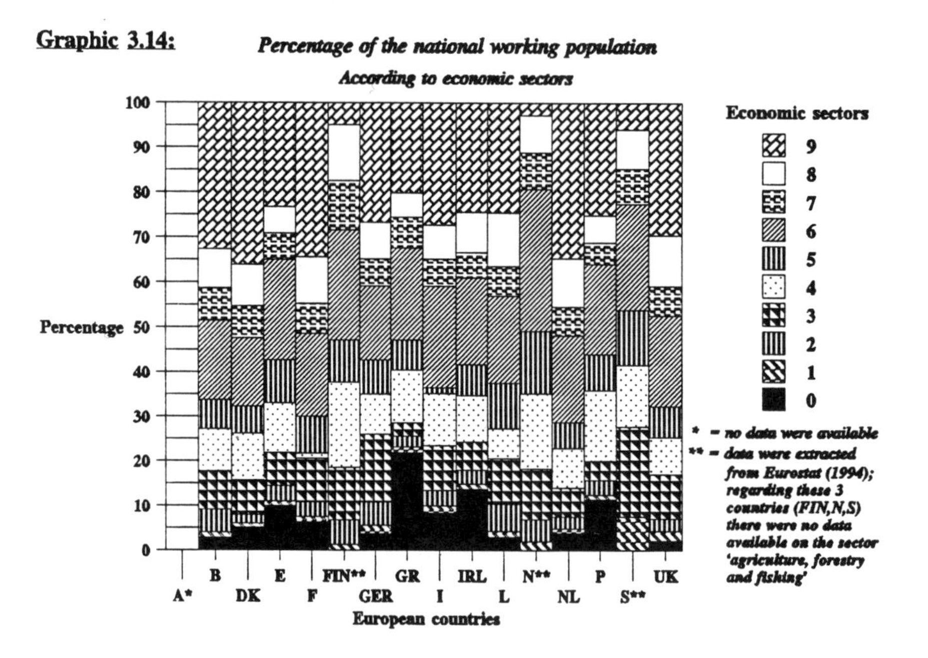

Economic sectors:
9 = Other services;
8 = Banking and finance, insurance, business services, renting;
7 = Transport and communications;
6 = Distributive trades, hotels, catering, repairs;
5 = Building and civil engineering;
4 = Other manufacturing industries;
3 = Metal manufacturing, mechanical and electrical industry;
2 = Extraction and processing of non-energy producing minerals, chemical industries;
1 = Energy and water;
0 = Agriculture, forestry and fishing.

Source: Eurostat, 1992a/1994

Graphic 3.14 shows the percentage of the national working population according to economic sectors (in 1992).

In eleven European countries (B,DK,E,F,GER,I,IRL,L,NL,P,UK) the largest part of the national working population are working in the *'other services'* sector (economic sector nr. 9). The range of percentage varies from circa 23%(E) to 36%(DK) of the national working population. In Finland, Norway and Sweden the economic sector *'distributive trades'* has the highest percentage, respectively circa 25%, circa 32% and circa 24% of the national working population. Finally, in Greece most workers are working in the *'agriculture, forestry and fishing'* sector: circa 22% of the national working population.

4 *Availability and quality of data on occupational health and safety in Europe*

Introduction

This chapter will present tables with data from national sources, i.e. ***sources other than the First European Working Environment Survey 1991-1992*** (European Foundation for the Improvement of Living and Working Conditions, 1992). In total the data of sixteen national reports have been taken up into the consolidated report.

Regarding availability of data on risk factors of the working environment six countries (Greece, Ireland, Italy, Luxemburg, Portugal and United Kingdom) only provided information from the Working Environment Eurosurvey mentioned above. For that reason, their country codes will not be present in the tables of this chapter which regard the working environment.

Firstly the availability of data and the data sources used in the various countries is dealt with (section 4.1). Then the quality of the data in terms of reliability and comparability will be discussed but only for the data which are best available. This information is taken up in section 4.2. This focus on best available data will be continued in chapter 5 as well, where the actual figures will be presented, along with qualitative interpretations.

4.1 Availability of data and sources of information

In this section an overview is presented of the available data on working environment and health and safety output (section 4.1.1). Thereafter, an overview is presented of the kinds of information sources present in the various countries, which were used for the national data gathering and reporting (section 4.1.2).

4.1.1 Availability of data

The presentation of data on 'working environment' and on 'health and safety output' hereafter distinguishes two categories:

1) Best available data: data from national sources, available in ≥ 6 countries in total;
2) Least available data: data from national sources, available in less than 6 countries.

For both categories tables and descriptions are taken up in this chapter, which give an overview of the availability of data across Europe (best available data: tables 4.1.1 and 4.1.2; least available data: tables 4.1.3 and 4.1.4). This regards not only the availability of data on the national working populations, but also of data according to gender, age, occupation and economic sector.
The categories of the variables occupation, economic sector and age, that were included in the matrix, are taken up in the annexes (3,4 and 7). Furthermore, the list of risk factors (exposures) used in the matrix is taken up in Annex 5. In Annex 6 a list of diagnostic categories of diseases (ICD, version 10) has been included. Insofar in this report is refered to the matrix, than this refers to the categories and definitions included in the respective annexes.
Furthermore, it should be stressed here, that the term 'available' means that national reports actually provided data on a specific topic. 'Unavailable' means that national reports contained no information on that topic (blanks), or had the remark 'no data available'. This may have several backgrounds, however. Data may not exist at all, or data may exist, but not according to the matrix, and/or not accessible or convertable within the time restrictions of the project.

Table 4.1.1a: *Available data of national sources on risk factors of the working environment (in ≥ 6 countries in total)*

Issue	Number of countries	Countries[1]	Data on gender N: Countries	Data on age N: Countries	Data on occupation N: Countries	Data on economic sector N: Countries
Physical exposure:	**10**					
▸ Noise	10	A;B;DK;E;F;FIN;GER;N;NL;S	6: A;DK;FIN;N;NL;S	4: DK[6];FIN[1];N[2];S[3]	3: A[4];GER[5];N[2]	5: DK[6];GER[5];N[2]; NL;S[3]
▸ Vibrations	9	A;B;DK;E;F;FIN;N;NL;S	6: A;DK;FIN;N;NL;S	4: DK[6];FIN[1];N[2];S[3]	2: A[4];N[2]	5: A[4];DK[6];N[2];NL;S[3]
▸ Climate	10	A;B;DK;E;F;FIN;GER;N;NL;S	6: A;DK;FIN;N;NL;S	3: DK[6];FIN[1];S[3]	3: A[4];GER[5];N[2]	5: DK[6];GER[5];N[2];NL;S[3]
▸ Radiation	6	B;DK;F;FIN;NL;S	3: DK;FIN;S	3: DK[6];FIN[1];S[3]	1: N[2]	3: DK[6];NL[2];S[3]
Psychosocial exposure:	**8**					
▸ Working pace	6	A;DK;E;GER;N;NL	4: A;DK;N;NL	2: DK[6];N[2]	1: A[4]	5: A[4];DK[6];GER[5];N[2];NL
▸ Job content	8	A;DK;E;FIN;GER;N;NL;S	6: A;DK;FIN;N;NL;S	4: DK[6];FIN[1];N[2];S[3]	2: A[4];N[2]	6: A[4];DK[6];GER[5];N[2];NL;S[3]
▸ Working hours	8	A;B;DK;E;GER;N;NL;S	6: A;B;DK;N;NL;S	5: B[7];DK[6];N[2];NL[8];S[3]	4: A[4];GER[5];N[2];S[3]	7: A[4];B[7];DK[6];GER[5];N[2];NL;S[3]
▸ Influence and control	7	DK;E;FIN;GER;N;NL;S	4: DK;FIN;N;S	4: DK[6];FIN[1];N[2];S[3]	2: N[2];S[3]	4: DK[6];GER[5];N[2];S[3]
▸ Social interaction	7	A;DK;E;FIN;GER;N;S	6: A;DK;FIN;N;NL;S	4: DK[6];FIN[1];N[2];S[3]	4: A[4];DK[6];N[2];S[3]	5: A[4];DK[6];GER[5];N[2];S[3]
Physiological exposure:	**8**					
▸ Working postures and movements	7	DK;E;FIN;GER;N;NL;S	5: DK;FIN;N;NL;S	4: DK[6];FIN[1];N[2];S[3]	1: GER[5]	3: DK[6];N[2];S[3]
▸ Manual lifting, handling, pushing and pulling	8	A;DK;E;FIN;GER;N;NL;S	6: A;DK;FIN;N;NL;S	4: DK[6];FIN[1];N[2];S[3]	2: A[4]; GER[5]	4: A[4];DK[6];N[2];S[3]

Remarks:

1) FIN: the different age categories do not correspond with the age categories of the matrix. The age categories in the Finnish report: (1) 15-24 years; (2) 25-49 years and (3) 50-64 years.
2) N: the categories on occupation do not correspond with the age-, occupation and economic sector categories of the matrix. Also different age catgories: (1) 16-24 years; (2)25-44 years; (3) 45-54 years and (4) 55-66 years.
3) S: different age categories are taken up into the report: (1) 16-24 years; (2) 25-49 years and (3) 50-64 years.
Also the data on economic sectors do not correspond with the occupation and economic categories of the matrix.
4) A: the categories of occupation do not correspond with the categories of the matrix. The categories for economic sectors refer only to the self-employed and helpers; a distinction is made between agriculture and forestry versus unspecified other sectors.
5)GER: the categories on occupation and branches of industries do not correspond with the occupation and economic sector categories of the matrix.
6) DK: different age categories are used: (1) below 18 years; (2) 18-24 years; (3) 25-49 years; (4) 50-64 years and (5) 65 years or more. These age categories do not correspond with the categories of the matrix. Also the economic sector categories used in the report do not correspond with the economic sector categories of the matrix.
7) B: different age categories are used: (1) 15-24 years; (2) 25-34 years; (3) 35-44 years; (4) 45-54 years and (5) 55 years or more. For the risk factor 'working hours' the categories used on economic sectors do not correspond entirely with the categories of the matrix.
8) NL: age categories used: (1) 15-24 years; (2) 25-44 years; (3) 45-54 years and (4) 55-64 years.

Table 4.1.1a shows that, regarding **risk factors of the working environment**, there are data from national sources available in 6 or more countries on the following issues:

- physical exposure: noise, vibrations, climate and radiation;
- psychosocial exposure: working pace, job content, working hours, influence and control over own work and social interaction;
- physiological exposure: working postures and movements and manual lifting, handling, pushing and pulling.

Physical exposure:

The categories on age, occupation and economic sector, provided by the countries, generally do not correspond with the categories of the matrix (see remarks above).

1. **Noise:**
Ten countries have provided data on this physical risk factor. Seven countries (A,DK,F, FIN,N,NL,S) also have provided data on noise according to *gender*.
Regarding data on noise five countries (DK,F,FIN,N,S) have provided information on *age*.
Three countries (A,GER,N) have provided data on *occupation*.
Finally, five countries (DK,GER,N,NL,S) have provided data according to *economic sector*.

2. **Vibrations:**
   Nine countries (A,B,DK,E,F,N,NL,S) have provided data on this risk factor. Six countries (A,DK,FIN,N,NL,S) also have provided data on vibrations according to *gender*. Two countries (A,N) have data on *occupation*.
   Finally, five countries (A,DK,N,NL,S) have data on vibrations according to *economic sector*.

3. **Climate:**
   Ten countries (A,B,DK,E,F,GER,N,NL,S) have provided data. Six of them (A,DK, FIN,N,NL,S) have provided data according to *gender*. Three countries (DK,FIN,S) provided figures according to *age*.
   Also three countries (A,GER,N) have provided data on *occupation*. Five countries (DK, GER,N,NL,S) have provided data on *economic sector*.

4. **Radiation:**
   Six countries (B,DK,F,FIN,NL,S) have provided data on radiation. Three countries (DK,FIN,S) have data on exposure to radiation of male and female workers (*gender*).
   The same three countries have also provided data according to *age*. Only one country (N) provided data on *occupation*. Finally, three countries (DK,NL,S) have data on *economic sector*.

Psychosocial exposure:

The categories on age, occupation and economic sector, provided by the countries, generally do not correspond with the categories of the matrix (see remarks beneath table 4.1.1a).

1. **Working pace:**
   Six countries (A,DK,E,GER,N,NL) have provided data on this psychosocial risk factor. Four of them (A,DK,N,NL) have data according to *gender*. Two countries (A,DK) have provided data on working pace according to *age* and only one country (A) has data on *occupation*. Finally, five countries (A,DK,GER,N,NL) have data on working pace according to *economic sector*.

2. **Job content:**
   Eight countries (A,DK,E,FIN,GER,N,NL,S) have provided data on job content as a psychosocial risk factor. Six countries (A,DK,FIN,N,NL,S) have data on exposure to job content of male and female workers (*gender*). Four countries (DK,FIN,N,S) have data on *age*-categories and two countries (A,N) on categories of *occupation*. Finally, six countries (A,DK,GER,N,NL,S) have provided data on *economic sector*.

3. **Working hours:**
   Eight countries (A,B,DK,E,GER,N,NL,S) have provided data. Six countries (A,B,DK, N,NL,S) have data on *gender*. Five countries (B,DK,N,NL,S) have data on working hours according to *age*. Categories on *occupation* are provided by four countries (A, GER,N,S) and finally, seven countries have data on *economic sector*.

4. **Influence and control over own work:**
Seven countries (DK,E,FIN,GER,N,NL,S) have provided data on this psychosocial risk factor. Four countries (DK,FIN,N,S) have provided data on *gender* and *age*. Two countries (N,S) have data on *occupation* and finally, four countries (DK,GER,N,S) have provided figures on *economic sector*.

5. **Social interaction:**
Seven countries (A,DK,E,FIN,GER,N,S) have provided data on social interaction as a psychosocial risk factor. Six countries (A,DK,FIN,N,NL,S) have data on exposure to social interaction of male and female workers (*gender*). Four countries (DK,FIN,N,S) have information on figures of social interaction according to *age*. Four countries (A,DK,-N,S) have provided data on *occupation* as well as *economic sector*.

Physiological exposure:

The categories on age, occupation and economic sector, provided by the countries, generally do not correspond with the categories of the matrix (see also remarks beneath table 4.1.1a).

1. **Working postures and movements:**
Seven countries (DK,E,FIN,GER,N,NL,S) have provided data. Five countries (DK,FIN, N,NL,S) have data according to *gender* and *age* (except NL). One country (GER) has provided information on figures regarding *occupation*. Finally, four countries (DK, GER,N,S) have data on *economic sector*.

2. **Manual lifting, handling, pushing and pulling:**
Eight countries (A,DK,E,FIN,GER,N,NL,S) have provided data on this psychosocial risk factor. Six countries (A,DK,FIN,N,NL,S) have provided information on figures of exposure according to *gender*. Four countries (DK,FIN,N,S) have data on *age* and two countries (A,GER) on *occupation*. Finally, four countries (A,DK,N,S) have provided data on *economic sector*.

Table 4.1.1b: *Availability of data on health and safety output from national sources (in ≥ 6 countries in total)*

Issue	Number of countries	Countries	Data on gender N: Countries	Data on age N: Countries	Data on occupation N: Countries	Data on economic sector N: Countries
Reported occupational accidents:	**15**					
▸ All accidents	10	A;B;E;F;FIN;;GR; IRL;L;P;UK	4: A;B;L[8];UK	5: A[1];B;F;L[8];UK	3: A[1];B[2];F	4: A[1];B[2];GR;IRL[4]
▸ Fatal accidents	15	A;B;DK;E;F;FIN;GER;GR; I;IRL;L;NL;P;S;UK	8: B;DK;E;GER[7];I;S;UK; NL	5: B;DK;E;GER[7];I	3: B[2];DK;GER[7]	7: B[2];DK[4];E;GER[7];I; IRL[4];S[4]
▸ With work interruption	11	B;DK;E;F;FIN;GER;I;IRL; NL;S;UK	9: B;DK;E;F;FIN;GER[7];I; NL;S	8: B;DK;E;F;FIN;GER[7]; I;NL	4: B[2];DK;F;GER[3]	7: B[2];DK[4];E;GER[7];I; NL;S[4]
▸ Accidents according to causes:	11	A;B;DK;E;FIN;GER;IRL; L;NL;S;UK	5: A;DK;NL;GER[7];S	3: A[1];DK;GER[7]	2: DK;GER[7]	7: B[2];DK[4];E;GER[7];L; S[4];UK
Sickness absence in general	**10**					
All cases:	10	A;B;DK;F;FIN;GER;L;NL; S;UK	9: A;B;DK;F;FIN;GER; NL;S;UK	7: B;FIN;GER;L;NL;S;UK		2: FIN;GER[7]
Reported occupational diseases:	**12**					
▸ Total of reported occupa-tional diseases	12	A;B;DK;E;F;FIN;GER;I; IRL;L;NL;S	7: B;DK;E;FIN;GER[7];I:S	6: B[5];DK;E[5];FIN[5];GER[7];I	5: B[5];DK;E[5];FIN[5];GER[7]	5: A[1];DK[6];E;GER[7];I
▸ Cases according to diagnosis:	12	A;B;DK;E;F;FIN;GER; I;IRL;L;S;UK	5: DK;E;FIN;GER[7];S	5: DK;E[5];FIN[5];GER[7];I	5: B[5];DK;E[5];FIN[5];GER[7]	4: A[1];DK[6];E;GER[7]
Occupational mortality	**9**					
▸ All cases	9	A;B;DK;F;FIN;I;NL;S;UK	5: DK;FIN;I;NL;UK	1: UK	1: UK	
General mortality	**12**					
▸ All cases	12	A;B;DK;E;FIN;GER[7];GR; I;L;NL;S;UK	10: A;DK;E;FIN;GER[7]; GR;I;L;NL;S	9: A;B;DK;FIN;GER[7]; GR;I;L;S		
▸ Cases according to diagnosis:	8	A;B;E;FIN;GER[7];GR;I;NL	6: A;E;FIN;GER[7];I;NL	7: A;B;E;FIN;GER[7];GR;NL		

Remarks:

1) <u>A:</u> the categories on age, occupation and economical sector do not correspond with the categories of the matrix. Age categories used: (1) ≤ 19 years; (2) 20-29 years; (3) 30-39 years; (4) 40-49 years; (5) 50-59 years; (6) 60-69 years; (7) ≥ 70 years.

2) <u>B:</u> the categories on age, occupation and economical sector do not correspond with the categories of the matrix. Data on occupation can only be provided according to BIT-classification. Age categories used: (1) ≤ 14 years; (2) 15-24 years; (3) 25-49 years; (4) 50-64 years; (5) ≥ 65 years.

3)<u>GER:</u> the categories on age, occupation and economical sector do not correspond with the categories of the matrix. Age categories used: (1) ≤ 19 years; (2) 20-24 years; (3) 25-29 years; (4) 30-34 years; (5) 35-39 years; (6) 40-44 years; (7) 45-49 years; (8) 50-54 years; (9) 55-59 years; (10) 60-64 years; (11) ≥ 65 years.

4) for these three countries (IRL,DK,S) the categories on economical sector do not correspond with the categories of the matrix.

5) for these 3 countries (B,E,FIN) the categories on age and occupation do not correspond with the categories of the matrix. Age categories used by:
Belgium: (1) ≤ 19 years; (2) 20-24 years; (3) 25-29 years; (4) 30-34 years; (5) 35-39 years; (6) 40-44 years; (7) 45-49 years; (8) 50-54 years; (9) 55-59 years; (10) 60-64 years; (11) ≥ 65 years.
Spain: (1) 15-24 years; (2) 25-34 years; (3) 35-44 years; (4) 45-54 years; (5) 55-64 years; (6) ≥ 64 years.
Finland: (1) 15-24 years; (2) 25-49 years; (3) 55-59 years; (4) 60-65 years.
Occupational categories used by: Belgium: ISCO-classification 1958 and Spain: CNO-classification.

6) for Denmark the categories on economical sectors do not correspond with the categories of the matrix.

7)<u>GER:</u> although data were not available from the national report, these country codes were added to this table on recommendation of the Steering Group member from the Bundesanstalt für Arbeitsschutz.

8)<u>L:</u> data according to gender and age are only available for agriculture and forestry.

Table 4.1.1b shows that, regarding **health and safety output**, there are data from national sources available in 6 or more countries on the following topics:

- occupational accidents: all accidents, fatal accidents, accidents with work interruption and causes of accidents;
- sickness absenteeism in general: all cases (lost days, total of cases, volume, frequency and average duration of spell);
- reported occupational diseases: total of reported occupational diseases and causes of reported occupational diseases);
- occupational mortality: all cases;
- general mortality: all cases and causes of general mortality.

<u>**Occupational accidents:**</u>

The categories on age, occupation and economic sector, provided by the countries, generally do not correspond with the categories of the matrix (see remarks beneath table 4.1.1b).

1. **All accidents:**
Ten countries (A,B,E,F,FIN,GR,IRL,L,P,UK) have provided data. Only four countries (B,L,NL,UK) have information on all occupational accidents according to *gender*. Six countries (A,B,F,L,NL,UK) have provided data according to *age*. Austria(A), Belgium(B) and France(F) have provided information on all accidents according to *occupation*. Finally, four countries (A,B,GR,IRL) have provided data according to *economic sector*.

2. **Fatal accidents:**
Fifteen countries (A,B,DK,E,F,FIN,GER,GR,I,IRL,L,NL,P,S,UK) have provided data on fatal accidents. Seven countries (B,DK,E,GER,I,S,UK) have taken up information of male and female workers (*gender*) who were exposed to this issue. Five countries (B,DK,E, GER,I) have data on *age*. Only three countries (B,DK,GER) have data on occupation and seven countries (B,DK,E,GER,I,IRL,S) on *economic sector*.

3. **Accidents with work interruption:**
Eleven countries (B,DK,E,F,FIN,GER,I,IRL,NL,S,UK) have provided data on this issue. Nine countries (B,DK,E,F,FIN,GER,I,NL,S) have provided figures on *gender* and seven countries (B,DK,E,F,FIN,GER,I) on *age*. Regarding *occupational* categories four countries (B,DK,F,GER) have provided data and seven countries (B,DK,E,GER,I,NL,S) have data on 'occupational accidents work interruption' according to *economic sector*.

4. **Causes of accidents:**
Eleven countries (A,B,DK,E,FIN,GER,IRL,L,NL,S,UK) have provided data. Five countries (A,DK,GER,NL,S) have data on causes of accidents according to *gender*. Austria, Denmark and Germany are the only countries that has provided data on *age*. Two countries (DK,GER) have data on *occupation*. Finally, seven countries (B,DK,E,GER,L,-S,UK) have data on *economic sector*.

Sickness absenteeism in general:

1. **All cases:**
Ten countries (A,B,DK,F,FIN,GER,L,NL,S,UK) have provided data regarding one of the following parameters of sickness absenteeism (lost days, total number of cases, volume, frequency and average duration of spell). Nine countries (A,B,DK,F,FIN,GER,NL,S,UK) have data on *gender*. Seven countries (B,DK,GER,L,NL,S,UK) have provided *age*-categories. There is no data available on *occupation* and only two countries (FIN,GER) have provided data on *economic sector*.

2. **Cases according to diagnosis:**
Regarding this category there are not enough data available (i.e. in less than 6 countries). Therefore this category has not been taken up in table 4.1.1b.

Reported occupational diseases:

1. **Total of reported occupational diseases:**
Twelve countries (A,B,DK,E,F,FIN,GER,I,IRL,L,NL,S) have provided data. Seven countries (B,DK,E,I,FIN,GER,S) have data on *gender*. Six countries (B,DK,E,FIN, GER,I) have provided data on *age* and five (B,DK,E,FIN,GER) on *occupation*. Finally, five countries (A,DK,E,GER,I) have data on *economic sector*.

2. **Cases according to diagnosis:**
Twelve countries (A,B,DK,E,F,FIN,GER,I,IRL,L,S,UK) have provided data. Five countries (DK,E,FIN,GER,S) have also information on *gender*. Five countries (DK.E, FIN,GER,I) have provided data on *age*-categories. Five countries (B,DK,E,FIN,GER) have provided figures on *occupations* and four countries (A,DK,E,GER) have done that on *economic sectors*.

3. **Notified occupational diseases and cases of notified occupational diseases:**
Regarding these two categories there are not enough data available (i.e. in less than 6 countries). Therefore this category has not been taken up in table 4.1.1b.

Occupational mortality:

Regarding this issue some overlap exists in some countries with 'Fatal occupational accidents'. In chapter 5 table 5.2.4 this problem will be dealt with.

1. **All cases:**
Nine countries (A,B,DK,F,FIN,I,NL,S,UK) have data on this issue. Five countries (DK, FIN,I,NL,S) have provided some figures on *gender*. Only the United Kingdom(UK) have some data on *age* and *occupation*. There are no data available on *economic sector*.

2. **Cases according to diagnosis:**
Regarding this category there are not enough data available (i.e. in less than 6 countries). Therefore this category has not been taken up in table 4.1.1b.

General mortality:

1. **All cases:**
Twelve countries (A,B,DK,E,FIN,GER,GR,I,L,NL,S,UK) have provided data on total number of cases. Ten of them (A,DK,E,FIN,GER,GR,I,L,NL,S) have some more figures on *gender*. Finally, nine countries (A,B,DK,FIN,GER,GR,I,L,S) have provided data on *age*. There are no data available on *occupation* and *economic sector*.

2. **Diagnoses related to general mortality:**
Eight countries (A,B,E,FIN,GER,GR,I,NL) have provided data on this issue. Six countries (A,E,FIN,GER,I,NL) have some more figures available on *gender* and seven countries (A,B,E,FIN,GER,GR,NL) on *age*-categories. There are no data available on *occupation* and *economic sector*.

Table 4.1.1c: ***Availability of data on the working environment from national sources in less than 6 countries***

Issue	Range[1] of countries	Countries	Data on gender N: Countries	Data on age N: Countries	Data on occupation N: Countries	Data on economic sector N: Countries
Physical exposure: lighting; pressure	3-5	A;B;DK;E;F;N;S	5: A;DK;FIN;N;S	3: DK;FIN;N;S	2: A[3];N	4: A[3];DK;N;S
Chemical exposure: (specific chemical agents; materials and compounds; products; dust, vapours, fumes)	3-7[2]	A;B;DK;F;FIN;GER;S	5: A;DK;F;NL;S	5: A;DK;F;NL;S	2: A;GER	5: A[3];DK;GER; NL;S
Biological exposure: (viruses; bacteria; growths; animals;foodstuff; vegetable fibers, wood, wood products; other material from biological origin; biochemical materials; materials from human beings	0-3	B;DK;F;FIN	2: DK;F	2: DK;F	1: F -	1: DK
Psychosocial exposure: (form of payment, traumatic experience)	4-5	DK;FIN;N;NL;S	5: DK;FIN;N;NL;S	4: DK;FIN;N;S	1: N	3: DK;N;S
Physiological exposure: (use and handling of tools and equipment	3	A;E;N	2: A;N	1: N	1: A	2: A[3];N
Exposure to safety risks: (falling to lower level; falling at the same level; getting struck by moving objects; getting trapped/pinched/crushed; getting cut/stabbed; horizontal and vertical transport; fire and explosions; electroshocks and electrocutions; con-tact with hot/cold objects)	0-5	A[4];E;FIN;GER[5];NL	2: A;NL		2: A[4];GER[5]	3: A[3,4];E;NL
Other exposures	1-1	A;DK;F;NL;S	5: A;DK;F;NL;S	4: DK;F;NL;S	1: A	3: A[3];DK;S

Remarks:

1) The range of countries refers to the maximum number of countries on sub-items of each risk factor.
2) Although in total 7 countries provided data on chemical exposures, these data have not been included in table 4.1.1a due to the ambiguity of the provided information.
3) Refers only to the self-employed and helpers; a distinction is made between Agriculture and Forestry versus unspecified other sectors.
4) Refers to dangers of accidents or injuries.
5) Risks not specified.

Table 4.1.1c shows that, regarding **risk factors of the working environment**, there are data from national sources available in less than 6 countries on the following topics:

- physical exposure: lighting and pressure;
- chemical exposure: specific chemical agents, materials and compounds, products and dust, vapours and fumes;
- biological exposure: viruses, bacteria, growths, animals, foodstuff, vegetable fibers, wood, wood products, other material from biological origin, biochemical materials and materials from human beings;
- psychosocial exposure: form of payment and traumatic experience;
- physiological exposure: use and handling of tools and equipment;
- exposure to safety risks: falling to lower level; falling at the same level; getting struck by moving objects; getting trapped/pinched/crushed; getting cut/stabbed; horizontal and vertical transport; fire and explosions; electroshocks and electrocutions; contact with hot/cold objects;
- other exposures: insufficient protective clothing, insufficient decoration of the working place, tobacco smoke more than seldom, dirty work, stench at work, display monitor.

General remark:

The categories on age, occupation and economic sector, provided by the countries, generally do not correspond with the categories of the matrix (see also remarks beneath table 4.1.1a). Regarding data on age and economic sector only The Netherlands(NL) uses categories that correspond with the categories of the matrix.

Physical exposure:

1. **Lighting:**
   Six countries (A,DK,E,N,S) have provided data on lighting. All six countries have data on male and female workers (*gender*). Three countries (DK,N,S) have provided data on *age*, two countries (A,N) on *occupation* and four countries (A,DK,N,S) on *economic sector*.

2. **Pressure:**
   Only two countries (B,F) have provided data on pressure. There are no figures provided regarding *gender*, *age*, *occupation* or *economic sector*.

Chemical exposure:

1. **Specific chemical agents:**
   Six countries (A,B,DK,F,FIN,NL) have provided some information. There is no data available on *gender* and *age*. Austria has provided data on *occupation* and Denmark and the Netherlands have done that for *economic sector*.

2. **Materials and compounds:**
   Six countries (A,B,DK,FIN,GER,S) have provided data. Three countries (A,DK,S) have data on *gender* and *age*-categories. Two countries (A,GER) have figures on *occupation*. Four countries (A,DK,GER,S) have provided data on *economic sector*.

3. **Products:**
Five countries (A,B,DK,FIN,S) have provided data. Three countries (A,DK,S) have data on *gender*, *age*-categories and *economic sector*. Only Austria has provided data on *occupation*.

4. **Dust, vapours, fumes:**
Seven countries (A,B,DK,F,FIN,GER,S) have provided data. Five countries (A,DK,F, NL,S) have provided figures on exposure to this issue by male and female workers (*gender*). Four countries (DK,F,NL,S) have done that on *age*, two countries (A,GER) on *occupation* and three countries (DK,GER,S) on *economic sector*.

It should however be noted that on the subject of chemical exposure the data appear to be quite ambiguous. Therefore these data - especially with regard to 'dust, vapours and fumes' have not been taken up in table 4.1.1a and consequently have not been taken up in the overview of reliability and comparability.

Biological exposure:

1. **Viruses:**
Only France has indicated that there are data available, including data on *age*-categories and *occupation*. These data refer to both viruses and bacteria.

2. **Bacteria:**
Three countries (B,F,FIN) have provided data. Only France has indicated that there is information available on *age*-categories and *occupation*.

3. **Growhts, animals, foodstuff, vegetable fibers etc., biochemical materials and materials from human beings:**
Regarding these six issues not one country has provided data.

4. **Other materials from biological origin:**
Only Denmark has provided data (on *gender*, *age* and *economic sector*) on this issue.

Psychosocial exposure:

1. **Form of payment:**
Five countries (DK,FIN,N,NL,S) have provided data on this issue. Five countries (DK,FIN,N,NL,S) have data on *gender*. Four countries (DK,FIN,N,S) have provide figures on *age* and three (DK,N,S) on *economic sector*. One country (N) has data on *occupation*.

2. **Traumatic experience:**
Four countries (DK,FIN,N,S) have provided data (on *gender* and *age*). Three countries (DK,N,S) have provided figures on *economic sector* and one country (N) on *occupation*.

<u>**Physiological exposure:**</u>

1. **Use and handling of tools and equipment:**
   Three countries (A,E,N) have provided data. Two countries (A,N) have data on *gender* and *economic sector*. One country (N) has provided data on *age*. Finally, Austria has provided data on *occupation*.

<u>**Exposure to safety risks:**</u>

Only five countries (A,E,FIN,GER;NL) have provided data (i.e. total number of exposed). One country (GER) provided data on *occupation* and unspecified safety risks. In total three countries (A,E,NL) have provided some more figures (on *gender, age, occupation* and *economic sector*) on the following issues:

1. **Falling to lower level, getting struck by moving objects, getting cut/stabbed, fire and explosions, electroshocks and electrocution and contact with hot/cold objects:**
   Two countries (A,NL) have provided data on *gender*, one country (A) on *occupation* and three countries (A,E,NL) on *economic sector*. There are no data available on *age*-categories.

2. **Falling at the same level, getting trapped/pinched/crushed and horizontal and vertical transport:**
   Two countries (A,NL) have provided data on *gender*, one country (A) on *occupation* and two countries (A,NL) on *economic sector*. There are no data available on *age*-categories.

<u>**Other exposures:**</u>

Five countries (A,DK,F,NL,S) have provided information on other exposures.

<u>Austria:</u> has provided information on: 'insufficient protective clothing' and 'insufficient decoration of the working place'. Regarding these issues this country has provided data on *gender*, *occupation* and *economic sector*.

<u>Denmark:</u> has provided information on 'tobacco smoke more than seldom'. Data are available on *gender*, *age* and *economic sector*.

<u>France:</u> has provided information on 'other exposures (in general)'. The data are available according to gender and age.

<u>The Netherlands:</u> has provided data on: 'dirty work' and 'stench at work'. Regarding these topics data is available on *gender* and *age* (categories do however not correspond with the categories of the matrix).

<u>Sweden:</u> has provided data on 'display monitor'. Data available according to *gender*, *age* and *economic sector*.

Table 4.1.1d: ***Availability of data on health and safety output from national sources in less than 6 countries***

Issue	Range of countries	Countries	Data on gender N: Countries	Data on age N: Countries	Data on occupation N: Countries	Data on economic sector N: Countries
Occupational sickness absenteeism: (all cases; cases according to diagnosis)	2-4	A;FIN;S;UK	1: S	-	-	1: FIN[1]
Occupational morbidity: (all cases; illness according to diagnosis)	1-3	E[2];S;UK[5]	1: S	1: S[3]	-	1: S[3]
General morbidity: (all cases; illness according to diagnosis)	3-4	E;FIN;GR;I	2: E;FIN	2: E[4];FIN[4]	-	-
Occupational disablement: (all cases; temporary; chronic; disability according to diagnosis)	1-3	B;E;F;I;NL	-	-	-	-
General disablement: (all cases; temporary; chronic; disability according to diagnosis)	0-5	B;DK;E;FIN;NL	5: B;DK;E;FIN;NL	5: B;DK;E;FIN;NL	-	-

Remarks:

1) FIN: the categories on economic sectors do not correspond with the economic sectors of the matrix.
2) E: 85% of the 9022 workers who visited health services proceed from companies of less than 6 workers and are not a representative sample of the industrial panorama of Spain, because they came from Other manufacturing industries (50%), Other services (22.5%), Metal manufacturing industries (8.8%) and Building industries (7.2%).
3) S: the categories on age and economical sectors don't correspond with the categories of the matrix.
4) for these two countries (E;FIN) different age categories (than those of the matrix) were taken up into the national report.
5) UK: these data refer to figures based on perceived work related ill-health as reported by the work force (1990, Labour Force survey Trailer Questionnaire).

Table 4.1.1d shows that, regarding **health and safety output**, there are data from national sources available in less than 6 countries on the following:

▸ occupational sickness absenteeism:	all cases, cases according to diagnosis;
▸ occupational morbidity:	all cases, illness according to diagnosis;
▸ general morbidity:	all cases, illness according to diagnosis
▸ occupational disablement:	all cases, temporary, chronic, disability according to diagnosis;
▸ general disablement:	all cases, temporary, chronic, disability according to diagnosis.

Occupational sickness absenteeism:

1. **All cases:**
   Four countries (A,FIN,S,UK) have provided data. Sweden has provided some figures on *gender* and Finland has done that on *economic sector*. Regarding *age* and *occupation* there is no data available.

2. **Cases according to diagnosis:**
   Two countries (S,UK) have provided data. Only Sweden has provided some figures on *gender*. Regarding *age*, *occupation* and *economic sector* there are no data available, with the exeption of Finland.

Occupational morbidity:

1. **All cases:**
   Three countries (E,S,UK) have provided data. Sweden is the only country that has provided data on *gender*.

2. **Illness according to diagnosis:**
   Only one country (S) has provided data (total number, according to *gender*, *age* and *economic sector*.

General morbidity:

1. **All cases:**
   Four countries (E,FIN,GR,I) have provided data. Two countries (E,FIN) have some more figures on *gender* and *age*-categories. There is no data available on *occupation* and *economic sector*.

2. **Illness according to diagnosis:**
   Three countries (E,FIN,GR) have provided data. Two countries (E,FIN) have some more figures on *gender* and *age*-categories. There is no data available on *occupation* and *economic sector*.

<u>**Occupational disablement:**</u>

For figures on this subject two countries (B,F) refer to occupational accidents and occupational diseases. Regarding *gender*, *age*, *occupation* and *economic sector* there is no data available.

1. **All cases:**
   Two countries (B,E) have provided data.

2. **Temporary occupational disablement and disabilitiy according to diagnosis:**
   Only one country (B) has provided data.

3. **Chronic occupational disablement:**
   Two countries (B,I) have provided data.

<u>**General disablement:**</u>

1. **All cases:**
   Four countries (DK,E,FIN,NL) have provided data. All four countries have some figures on *gender* and *age*. There is no data available on *occupation* and *economic sector*.

2. **Temporary and chronic occupational disablement:**
   There is no data available.

3. **Disability according to diagnosis:**
   Three countries (DK,FIN,NL) have provided data. Only Finland has provided figures on *gender*. All three countries have figures on *age*-categories. There is no data available on *occupation* and *economic sector*.

4.1.2 Sources of information

On the basis of available data in the national reports an overview of sources of information (table 4.1.2a and 4.1.2b) on working environment and health and safety output has been composed. It is likely that there are more sources existing than those which have been taken up in these tables. Therefore the tables below should not be considered to give an exhaustive and reliable overview of all existing sources of information.

The working environment:

Table 4.1.2a: ***Overview of the sources of information on risk factors of the working environment***

Sources of information	Countries
SURVEYS:	
Working Environment Survey	
▸ National level	A[?];B[2];DK[3];E[3];F[1];FIN[2,3];GER[?];N[2];NL[3];S[2]
▸ European level	B[4];E[4];F[4];GR[4];I[4];L[4];NL[4];P[4];(DK[5];GER[5]; IRL[5];UK[5])
Labour Force Survey	
▸ National level	B[2];NL[2];UK[2]
▸ European level	NL
Periodical Occupational Health Examination	B[1];NL[3]
Other Surveys/studies	NL[2]

Notes: within the European countries the following organizations are responsible for the publication of surveys on risk factors of the working environment (which are mentioned above):
(1) the government; (2) the national bureau for statistics; (3) a national research institute; (4) an European research institute; (5) it is known to contain data as well for these four countries; (?) it is unknown who is responsible for publication.

Table 4.1.2a shows that most of the information sources on risk factors of the working environment concerns several kinds of surveys:

Working environment surveys:

Twelve countries (A,B,DK,E,FIN,GER,GR,I,N,NL,P,S) have provided data from this kind of information source; ten of them (A,B,DK,E,F,FIN,GER,N,NL,S) on national and six countries (B,E,GR,I,NL,P) on european level.

National level: in one country (F) the *government* was responsible for the publication of the survey; in four countries (B,FIN,N,S) the *national bureau of statisitcs* had taken the responsibility; in five countries (DK,E,FIN,GER,NL) it was in the hands of a *national research institute* and from one country (A) it is unkown who was responsible for publication.

European level: eight countries (B,E,F,GR,I,L,NL,P) provided figures that were extracted from the European survey on the working environment, published by the European Foundation for the Improvement of the Living and Working Conditions (1992). Furthermore this survey contains data on the working environment for four other countries (DK,GER,IRL, UK) as well, although these countries did not provide information from it.

Labour force survey:

Two countries (B,NL) have provided data on this kind of information source; Belgium on national level, whereas the Netherlands provided it on national as well as European level. In respect of the information source on national level the national statistical bureaux of Belgium and The Netherlands took the responsibility for publication.

Periodical Occupational Health Examination:

Only the Netherlands and Belgium have provided data from this kind of information source, on which a national research institute, respectively the government, took the responsibility for publication.

Other surveys/studies:

Only the Netherlands(NL) have reported some more information of other studies. Data was extracted from a survey done by the national bureau of statistics. It is likely that other relevant studies and research exist in other countries as well (see also section 1.1). However there are no reports mentioned as an information source.

Health and safety output:

Table 4.1.2b: ***Overview of the sources of information on health and safety output***

Information source / Type of organization	Statistics[1-4]	Reports[5-7]	Surveys[8-11]	Registers and databases
Government	B[1];DK[1];E[1];FIN[1];IRL[1];UK[1]	B[5,7];IRL[5];NL[5,6]; S[5];UK[5-7]	E[11];UK[9,11]	
National bureau of statistics	A[1];B[1,3];DK[1,3];E[1];FIN[1,4]; GR[1];I[1];IRL[3];L[1];NL[1];S[1];UK[1]		FIN[8];IRL[9]; S[9]	E
International bureau of statistics	GER[3];GR[3];I[3];P[3]			
Occupational insurance fund	B[1];GER[1];NL[1]	B[5];L[5]	-	-
Social security institution	A[1];B[3];F[1];FIN[1];GER[1];NL[1]	B[5];NL[5,6]	-	-
National research institute	DK[1,2]	F[6];FIN[6,7]	E[11];NL[11]	DK;FIN
European research institute	-	-	B[11];F[11]; GER[11];L[11]	-

Three information sources (statistics, reports and surveys) are specified as followed:
- Statistics: (1) general statistics ;(2) statistics regarding the working environment; (3) statistics regarding the labour force; (4) labour market statistics
- Reports: (5) annual report; (6) research report; (7) publication
- Surveys: (8) working environment survey; (9) national labour force survey; (10) International (European) Labour Force Survey; (11) other studies;

Table 4.1.2b shows that seven types of organizations are responsible for five different kinds of information sources on health and safety. It should be noted however, that 'registers and databases' seem not well-chosen categories, because they are not sufficiently distinguishable form the other categories. Statistics, for example, are often register-based and extracted from databases.

The following types of organizations were mentioned in various reports:

Government:

Goverments of eight countries (B,DK,E,FIN,IRL,NL,S,UK) are responsible for the publication of 3 kinds of information source:

1. **Statistics:**
   The goverment of six countries (B,DK,E,FIN,IRL,UK) is responsible for the publication of general statistics.

2. **Reports:**
   The goverment of five countries (B,IRL,NL,S,UK) is responsible for publication of the following reports: anual report (B,IRL,NL,S,UK); research report (NL) and publication (B,UK).

3. **Surveys:**
   The government of two countries (E,UK) is responsible for the publication of the following surveys: national labour force survey (UK) and other studies (E,UK).

National bureau of statistics:

Twelve countries (A,B,DK,E,FIN,GR,I,IRL,L,NL,S,UK) have extracted data from the following kinds of publications:

1. **Statistics:**
   Eleven countries (A,B,DK,E,FIN,GR,I,L,NL,S,UK) have extracted data from the following kind of statistics: general statistics (A,B,DK,E,FIN,GR,I,L,NL,S,UK); statistics regarding the labour force (B,DK) and labour market statistics (FIN).

2. **Surveys:**
   Three countries (FIN,IRL,S) have extracted data from the following kind of surveys: working environment survey (FIN) and national labour force survey (IRL,S).

3. **Databases:**
   Only Spain has extracted data from this kind of information source.

International bureau of statistics

Three countries (GR,I,P) have used information sources from the International Labour Office (ILO) in Geneva. This organization is responsible for the publication of labour force statistics. Germany has reported Eurostat statistics as an information source.

Occupational insurance fund:

Four countries (B,GER,L,NL) have extracted data from the following kind of publications of an occupational insurance fund:

1. **Statistics:**
Belgium, Germany and the Netherlands have extracted data from a publication with general statistics.

2. **Reports:**
Only Belgium and Luxembourg extracted data from an annual report.

Social security institution:

Six countries (A,B,F,FIN,GER,NL) have extracted information from the following kind of publications of a social security institution:

1. **Statistics:**
Six countries (A,B,F,FIN,GER,NL) have extracted data from a publication with general statistics. One country (B) extracted data from statistics regarding the labour force.

2. **Reports:**
Two countries (B,NL) have extracted data from the publication with the following kind of reports: annual reports (B,NL) and research reports (NL).

National research institute:

Five countries (DK,E,F,FIN,NL) have extracted data from the following kind of publication of a national research institute:

1. **Statistics:**
Only Denmark has extracted data from two kind of statistics: general statistics and labor force statistics.

2. **Reports:**
Only France and Finland have extracted data from two kind of reports: research reports and a publication.

3. **Surveys:**
The Netherlands and Spain have extracted data from other studies.

4. **Registers/databases:**
Denmark, Finland and Spain have extracted data from this information source.

European research institute:

Four countries (B,F,GER,L) have extracted data from a publication of an European research institute. Also eight other countries were involved in this study: DK,E,GR,I,IRL,NL,P and the UK. Nevertheless this publication was not mentioned in these eight national reports. Furthermore Germany refers to Eurostat statistics.

4.2 Reliability and comparability of data

Introduction

In the following paragraph the information on the national sources of data in the various European countries with respect to reliability and comparability will be presented. For the purpose of this overview of reliability and comparability the available information has been categorized. As far as reliability is concerned a division has been made in 2 categories:
**: in general this refers to data from samples considered representative for the whole working population or data from national databases/registers.
*: this concerns data with only limited or no representative value, e.g. case studies, local research or such a limited number of respondents that extrapolation is not possible.
Furthermore a distinction has been made between the origins of the available information, i.e. subjective versus objective. Either the information was obtained through questionnaire based survey research or self-report (subjective), or through actual (technical) measurement (objective, e.g. measurement of noise levels, or the registration of occupational accidents or sickness absence data).
Concerning comparability a distinction has been made between 2 categories:
++: this refers to data exactly corresponding with the definitions and categories in the matrix
+: this refers to data not corresponding with the definitions and categories in the matrix
In addition to this, the most recent years on which information could be provided have been included in the tables.

The aspects gender, age, occupation and economic sector have not been screened specifically with regard to reliability and comparability. As the data on these various categories are generally derived from the same source as the general national data, it has been assumed that the aspect of reliability is being taken into account. As far as comparability of these aspects is concerned, this will be discussed if relevant (i.e. only for the best available data in ≥ 6 countries in total) in the descriptions in chapter 5.

4.2.1 Reliability of data

It should be noted beforehand that where data are considered reliable, this refers specifically to the aspect of the coverage or representative value of the data, as stated in the national reports. Within the framework of this report it is obvious that the reliability of the data, in terms of the methodology used to acquire them, could not be subjected to screening. In addition it should be stressed that most data concerning the working environment in the various countries were collected through 'subjective' methods. Although in itself this does not constitute a reliability problem, it means however that it is not possible to supplement the data on the working environment from the Eurosurvey with 'hard' or 'objective' data. Furthermore there is a major threat to reliability in terms of underreporting in case of health and safety output data. Although not in every national report specifically mentioned, there is a widespread feeling among experts that in many countries, if not all, this is a major problem concerning most of the output data.
Other aspects of reliability are being dealt with in chapter 7.

Table 4.2.1a: ***Reliability of data of national sources on risk factors of the working environment (in $\geq$ 6 countries in total)***

Issue	Number of countries	Reliability[1] ** N: Countries	Reliability[1] * N: Countries	Objective[1] information N: Countries	Subjective[1] information N: Countries
Physical exposure:	**10**				
▸ Noise	10	8: B;DK;E;FIN;GER;N;NL;S	2: A;F	3: B;F;NL	7: A;DK;E;FIN;GER;N;S
▸ Vibrations	9	7: B;DK;E;FIN;N;NL;S	2: A;F	3: B;F;NL	6: A;DK;E;FIN;N;S
▸ Climate	10	8: B;DK;E;FIN;GER;N;NL;S	2: A;F	3: B;F;NL	7: A;DK;E;FIN;GER;N;S
▸ Radiation	6	4: B;FIN;NL;S	2: F;DK[2]	3: B;F;NL	3: DK;FIN;S
Psychosocial exposure:	**8**				
▸ Working pace	6	5: DK;E;GER;N;NL	1: A	-	6: A;DK;E;GER;N;NL
▸ Job content	8	7: DK;E;FIN;GER;N;NL;S	1: A	-	8: A;DK;E;FIN;GER;N;NL;S
▸ Working hours	8	7: B;DK;E;GER;N;NL;S	1: A	-	8: A;B;DK;E;GER;N;NL;S
▸ Influence and control	7	7: DK;E;FIN;GER;N;NL;S		-	7: DK;E;FIN;GER;N;NL;S
▸ Social interaction	7	6: DK;E;FIN;GER;N;S	1: A	-	7: A;DK;E;GER;N;NL;S
Physiological exposure:	**8**				
▸ Working postures and movements	7	7: DK;E;FIN;GER;N;NL;S		-	7: DK;E;FIN;GER;N;NL;S
▸ Manual lifting, handling, pushing and pulling	8	7: DK;E;FIN;GER;N;NL;S	1: A	-	8: A;DK;E;FIN;GER;N;NL;S

1) See 4.2: Introduction
2) Denmark reported that the data on ionizing radiation are ambigious.

Table 4.2.1b: ***Reliability of data of national sources on health and safety output (in ≥ 6 countries in total)***

Issue	Number of countries	Reliability[#] ** N: Countries	Reliability[#] * N: Countries	Objective[#] information N: Countries	Subjective[#] information N: Countries
Occupational accidents:	**15**				
▸ All accidents	10	10: A;B;E;F;FIN;GR;IRL;L;P;UK	-	10: A;B;E;F;FIN[4];GR;IRL;L;P;UK	-
▸ Fatal accidents	15	15: A;B;DK;E;F;FIN;GER;GR;I;IRL; L;NL;P;S;UK	-	15: A;B;DK;E;F;FIN;GER;GR;I;IRL; L;NL;P;S;UK	-
▸ With work interruption	11	9: B;E;F;FIN;GER;I;IRL;S;UK	2: DK[1];NL[7]	11: B;DK;E;F;FIN[4];GER;I;IRL;NL;S;UK[3]	-
▸ Accidents according to causes:	11	9: A;B;E;FIN;GER;IRL;L;S;UK	2: DK[1];NL[7]	11: A;B;DK;E;FIN[4];GER;IRL;L;NL;S;UK[3]	-
Sickness absence in general	**10**				
All cases	10	6: A;FIN;GER;L;NL;S	4: B;DK;F;UK	6: A;DK;FIN;GER;L;NL	4: B;F;S;UK
Reported occupational diseases:	**12**				
▸ Total of reported occupational diseases	12	8: A;B;DK;E;FIN[6];GER;I;S;	4:IRL[2];F[8];NL;L[3]	12: A;B;DK;E;F;FIN;GER;I;IRL;NL;L;S	-
▸ Cases according to diagnosis	12	9: A;B;DK;E;FIN[6];GER;I;S;UK	3:F[8];IRL[2];L[3]	10: A;B;DK;E;F;FIN[6];GER;I;IRL;S	2: FIN[6];UK[5]
Occupational mortality	**9**				
All cases	9	8: A;B;F;FIN;I;NL;S;UK	1: DK	9: A;B;DK;F;FIN;I;NL;S;UK	-
General mortality	**12**				
All cases	12	11: A;B;DK;FIN;GER;GR;I;L;NL;S;UK	1: E	12: A;B;DK;E;FIN;GER;GR;I;L;NL;S;UK	-
Cases according to diagnosis:	8	7: A;B;FIN;GER;GR;I;NL;	1: E	8: A;B;E;FIN;GER;GR;I;NL;	-

#) See 4.2: Introduction

Remarks:

1) <u>DK:</u> regarding 'occupational accidents' the data on 'work with interruption' and 'accidents according to:..' are not very reliable.
2) <u>IRL:</u> the data on 'reported occupational diseases' and 'cases according to:' are not reliable.
3) <u>L:</u> regarding 'reported occupational diseases' the data only cover agriculture, forestry and industries.
4) <u>FIN:</u> regarding 'occupational accidents' it is not clear whether the data on 'all accidents', 'with work interruption' and 'accidents according to:..' is subjective or objective information.
5) <u>UK:</u> uncertain basis of information.
6) <u>FIN:</u> the status of reliability of the data regarding 'occupational accidents' is not clear. It is also not clear whether the data refer to objective or subjective information. This also applies for the category 'cases according to:..' of the reported occupational diseases.
7) <u>NL:</u> the reliability is questionable, due to limited coverage and severe underreporting.
8) <u>F:</u> data cover only workers under the 'Régime Géneral' (15 large sectors).

Working environment
Table 4.2.1a shows that most countries (B,DK,E,FIN,GER,N,NL,S) have provided reliable data on risk factors regarding physical, psychosocial and physiological exposure. The data provided by Austria and France were considered not reliable, due to the fact that data for both countries is relatively outdated (Austria: '85, France; '86-'87) and the fact that the data in both countries seems not sufficiently representative. Remarkably the data provided on the three kinds of exposure concern mainly subjective information (worker's perceptions). Only three countries (B,F,NL) have provided more objective information on physical risk factors in the working environment (Periodical Occupational Health Examinations; survey by occupational physicians).

Health and safety output
Table 4.2.1b shows that, regarding the issues on health and safety output, most countries have provided reliable data. The category 'all cases of sickness absenteeism in general' shows a marked variation of reliability. Six countries (A,FIN,GER,L,NL,S) have provided reliable figures and four (B,DK,F,UK) have provided unreliable data or from unclear origin.
In contrast to the available data on the three groups of environmental risk factors, mentioned in table 4.2.1a, the information on health and safety output is mainly obtained from objective sources.

4.2.2 Comparability of data

Working environment
Table 4.2.2a shows an overview of the comparability of the data that are available in six or more countries on the working environment (see also table 4.1.1a). Regarding the risk factors of physical, psychosocial and physiological exposure itself, it is apparent that the categories and definitions used in the national reports correspond only to a very limited extent with the categories and definitions of the matrix (see annex 5). Only on 'noise' have data been provided from seven countries (A,DK,E,FIN,GER,NL,S) with the categories according to the guidelines of the matrix. Yet even within these seven countries minor differences can be observed concerning the specific definitions of this risk factor.
It is however possible that countries within the group with 'lesser comparability' have among themselves comparable data available. This could not be assessed due to time constraints.
Apart from the problems with regard to the risk factors themselves, it is also apparent that on the aspects age, occupation and economic sector, major differences in definitions and categories used can be noted. These latter problems will however be addressed in chapter 5, wherever descriptions are given of sufficiently available data.

Regarding the reference period the countries have provided data on the types of exposure of different periods:

- '92-'93: Spain, Norway and Sweden for all risk factors of the working environment; the Netherlands have provided data for three psychosocial ('working pace', 'job content' and 'working hours') and the two physiological risk factors.
- '90-91: Belgium, Denmark, Finland, Germany and the Netherlands have provided data from this reference period. Denmark, Finland and Germany provided data on all three kinds of exposure. The Netherlands provided data on physical and psychosocial exposure. Finally, Belgium provided data on the four physical risk factors.
- Before '90: Two countries (A,F) provided data that refer to the period before 1990. Austria provided data from 1985 for all three kinds of exposure. France only provided data from 1986/87 for the four physical risk factors.

Health and safety output

Table 4.2.2b shows that only for 'occupational accidents'('all accidents','fatal accidents' and 'accidents with work interruption') is there a substantial number of countries that use categories corresponding with the categories of the matrix. As far as occupational disease is concerned data prove to be comparable for four countries . For the remaining output issues comparability is virtually impossible.

Regarding the reference period most countries have provided data that refer to the period '92-'93.

Table 4.2.2a: ***Comparability of data of national sources on risk factors of the working environment (in ≥ 6 countries)***

Issue	Number of countries	Comparability (++)[1] N: Countries	Comparability (+)[1] N: Countries	'92 - '93[1] N: Countries	'90 - '91[1] N: Countries	Before '90[1] N: Countries
Physical exposure:	**10**					
▸ Noise	10	7: A;DK;E;FIN;GER;NL;S	3: B;F;N	3: E;N;S	5: B;DK;FIN;GER;NL;	2: A;F
▸ Vibrations	9	3: A;E;NL	6: B;DK;F;FIN;N;S	3: E;N;S	4: B;DK;FIN;NL	2: A;F
▸ Climate	10	2: GER;NL	8: A;B;DK;E;F;FIN;N;S	3: E;N;S	5: B;DK;FIN;GER;NL	2: A;F
▸ Radiation	6	3: FIN;NL;S	3: B;F;DK[2]	1: S	4: B;DK;FIN;NL	1: F
Psychosocial exposure:	**8**					
▸ Working pace	6	1: GER	5: A;DK;E;N;NL	3: E;N;NL	2: DK;GER	1: A
▸ Job content	8		8: A;DK;E;FIN;GER;N;NL;S	4: E;N;S;NL	3: DK;FIN;GER	1: A
▸ Working hours	8		7: A;B;DK;E;GER;N;NL;S	5: B[3];E;N;S;NL	2: DK;GER	1: A
▸ Influence and control	7	2: DK;GER	5: E;FIN;N;NL;S	3: E;N;S	4: DK;FIN;GER;NL	
▸ Social interaction	7		7: A;DK;E;FIN;GER;N;S	3: E;N;S	3: DK;FIN;GER	1: A
Physiological exposure:	**8**					
▸ Working postures and movements	7		7: DK;E;FIN;GER;N;NL;S	4: E;N;NL;S	3: DK;FIN;GER	
▸ Manual lifting, handling, pushing and pulling	8		8: A;DK;E;FIN;GER;N;NL;S	4: E;N;NL;S	3: DK;FIN;GER	1: A

1) See 4.2: Introduction
2) Denmark reported that the data on ionizing radiation are ambigious.
3) Belgian data stem in fact from 1994.

Table 4.2.2b: ***Comparability of data of national sources on health and safety output (in ≥ 6 countries)***

Issue	Number of countries	Comparability ++[1] N: Countries	Comparability +[1] N: Countries	'92 - '93[1] N: Countries	'90 - '91[1] N: Countries	Before '90[1] N: Countries
Occupational accidents:	**15**					
▸ All accidents	11	10: A;B;E;F;FIN;GR;IRL;L;P;UK	1: GER	10: A;B;E;F;FIN;GER;IRL; L;P;UK	1: GR	
▸ Fatal accidents	15	15: A;B;DK;E;F;FIN;GER;GR;I; IRL;L;NL;P;S;UK		12: A;B;DK;E;F;FIN;GER; IRL;L;P;S;UK	2: GR;I	1: NL
▸ With work interruption	11	9: B;DK;E;F;FIN;I;IRL;NL;S	2: GER;UK	10: B;DK;E;F;FIN;GER; IRL;NL;S;UK	1: I	
▸ Accidents according to causes:	11	1: S	10:A;B;DK;E;FIN;GER; IRL;L;NL;UK	10: A;B;DK;E;FIN;GER IRL;L;S;UK		2: B;NL
Sickness absenteeism in general	**10**					
▸ All cases	10	1: NL	9: A;B;F;DK;FIN;GER;L;S;UK	8: A;B;F;GER;NL;S;L;UK	2: DK;FIN	
Reported occupational diseases:	**12**					
▸ Total of reported occupational diseases	12	4: B;E;FIN;IRL;	8: A;DK;F;GER;I;NL;L;S	11: A;B;DK;E;F;FIN;GER; IRL;NL;L;S		1: I
▸ Cases according to diagnosis:	12	1: E	11: A;B;DK;F;FIN;GER;I; IRL;L;S;UK	10: A;B;DK;E;F;FIN;GER; IRL;L;S;	1: UK	1: I
Occupational mortality	**9**					
▸ All cases	9	1: DK	8: A;B;F;FIN;I;NL;S;UK	7: A;B;F;FIN;I;S;UK		2: DK;NL
General mortality	**12**					
▸ All cases	12	2: FIN;UK	10: A;B;DK;E;GER;GR;I;L;NL;S	7: A;DK;FIN;GER;NL;S;UK	3: GR;I;L	2: B;E
Cases according to diagnosis:	8		8: A;B;E;FIN;GER;GR;I;NL	4: A;FIN;GER;NL	2: GR;I	2: B;E

1) See 4.2: Introduction

4.3 Discussion and preliminary conclusions

Based on the previously presented information from the national reports of 16 countries, and taking into account the limitations of this study due to its methodology, the following conclusions are drawn.

Conclusions on the availability of data in general:

- The overall picture emerges that data are generally more available across Europe with respect to health and safety output, than regarding the working environment. Moreover, data on occupational accidents appear to be the best available data of all (15 countries), followed by general mortality and occupational diseases (12 countries). Concerning the risk factors of the working environment, information on physical exposure is the most frequently available (10 countries). Data on other risk factors are only available in 8 countries or less.
- Hence, the focus on monitoring occupational health and safety across Europe so far appears to be mainly on the effects and health consequences of working conditions, and to a lesser degree on working conditions themselves which, from the point of view of prevention policy, would be preferable.

Conclusions on the availability of data on the working environment:

- In a substantial number to a majority of the countries (6-10 countries) data on various topics exist from information sources, other than the First European Survey on the Working Environment. These topics are:
  * physical exposure: noise, vibrations, climate, radiation;
  * psychosocial risk factors: working pace, job content, working hours, influence and control, social interaction;
  * physiological hazards: working postures, manual handling.
- In a minority of the countries (1-5 countries) data from information sources, other than the First European Survey on the Working Environment exist on other topics as well. These topics are:
  * physical exposure: lighting, pressure;
  * chemical exposure: chemical agents, materials and compounds, products, dust, vapours, fume;
  * biological exposure: e.g. viruses, bacteria, growths, animals, foodstuff, vegetable fibers, wood and products, biochemical material;
  * psychosocial exposure: form of payment, traumatic experiences;
  * physiological exposure: use and handling of tools and equipment;
  * exposure to safety hazards: falling, getting struck/trapped/pinched/crushed/cut/stabbed, transport, fire and explosions etcetera.
- The data on the working environment that are available in the various countries mostly concern the national working population as a whole. For practically all risk factors data according to gender, age and economic sector are only available in a slightly smaller number of countries. Data according to occupation are much less frequently available. On the whole, general data are slightly more available regarding risk factors in the psychosocial work environment, than with respect to the physical and physiological workload.

Conclusions on the availability of data on the health and safety output:

- In a substantial number to a majority of the countries (9-15 countries) data exist on the following topics:
  * occupational accidents;
  * sickness absenteeism in general;
  * occupational diseases;
  * occupational mortality;
  * general mortality.
- In a minority of the countries (1-5 countries) data exist on other topics as well:
  * occupational sickness absenteeism;
  * occupational morbidity;
  * general morbidity;
  * occupational disablement;
  * general disablement.
- The data on health and safety output that are available in the various countries mostly concern the working populations or national populations as a whole. For nearly all topics, data according to gender, age, occupation and economic sector are only available in a smaller number of countries. Exceptions to this rule are data on sickness absence in general, according to gender, and data on general disablement according to gender and age, which are present in practically all countries that provided data on these topics.
- On the whole, data according to gender, and age are generally available in more countries, than data according to occupation and economic sector, but per topic this pattern is rather diverse.

Conclusions on information sources:

- Besides the information from the first European Working Environment Survey (European Foundation for the Improvement of Living and Working Conditions, 1992) the provided data on the working environment in the various countries mostly stem from specific national questionnaire based surveys on the working environment, which are either separately held, or are attached to the Labour Force Survey as so-called trailer questionnaires (United Kingdom, Germany, Sweden, Denmark, France). In only two countries (Belgium and the Netherlands) the Labour Force Survey itself provides information on the working environment. Furthermore, in three countries (Belgium, France and the Netherlands) Periodical Occupational Health Examinations and/or specific studies form a source of information.
- The main providers of information on the working environment are generally the Eurosurvey from the Foundation (8 countries), national research institutes (5 countries) and the national bureaux of statistics (4 countries)
- Main information sources for data on health and safety output have appeared to be various sorts of statistics (14 countries). Surveys are in second place (10 countries). Reports such as annual or research reports form the third main source of information (7 countries), whereas registers and databases are the least common information sources (3 countries).
- The main providers of health and safety output data are apparently the national bureaux of statistics (12 countries) and governmental organisations (8 countries). Social security institutions, national research institutes and occupational insurance funds are the second main information providers (4-6 countries). Four countries provided information from internationally operating organisations, which are likely to have information on other countries as well however.

Conclusions on reliability of data:

- On the whole, data obtained on the working environment are mostly considered to be sufficiently reliable. For nearly all risk factors the data, provided by the majority of the countries, were judged sufficiently reliable. This means that the data either concern the entire working population, or are based on a representative sample which allows extrapolation towards the entire working population.
- Yet it should be noted that the data mostly concern subjective information, since they mostly stem from questionnaire-based surveys (see conclusions on information sources). Hence, the available data mainly regard worker's perceptions on exposure to risk factors in the working environment. Apart from Belgium, France and the Netherlands, as far as physical exposure is concerned, the data on the working environment are not 'confirmed' by 'objective' data.
- Reliability of data on health and safety output seems to be sufficient in most of the cases as well, at least as far as the coverage is concerned. However, in some countries under reporting was mentioned as a potential problem, especially in the field of occupational accidents and occupational diseases. Although not mentioned in all national reports the conclusion is justified, according to opinions of experts in the field, that under reporting, especially in the field of accidents and diseases, is a significant problem in all countries involved.
- Clearly the methodology by which data had been gathered, has a major influence on the reliability of data. In many cases however insufficient information was provided on the methodology aspect. Consequently this remains a blank spot in this analysis. This problem will be addressed further in chapter 7.
- Reliability concerning the aspects gender, age, occupation and economic sector has not specifically been screened in this report. It is assumed that while the data on these aspects generally have been derived from the same sources as the general data, the reliability in terms of coverage will be the same.

Conclusions on comparability of data:

- In general, regarding data on the entire working population the number of topics for which a substantial number of countries can provide comparable data is rather limited.
- The fact that for many topics various definitions and categories are used, severely hampers the comparability.
- Concerning risk factors of the working environment only on the issue 'noise' a substantial number (7) of countries was able to provide data more or less in accordance with the definition of the matrix. For the other risk factors most countries use deviating definitions or categories.
- Concerning health and safety output the comparability appears to be high for 'fatal accidents', (15 countries). To a lesser degree the data on 'all accidents','accidents with work interruption' and 'occupational diseases' are comparable (respectively for 10, 9 and 4 countries).
- Comparability on the issues gender, age, occupation and economic sector has not been dealt with in this chapter. These issues will be adressed if relevant in the descriptions of the actual data in chapter 5.

The overall conclusion on the availability, reliability and comparability of data on risk factors of the working environment and health and safety output indices, appear to be as follows:

- In 10 countries (A,B,DK,E,F,FIN,GER,N,NL,S) in this study data are available on one or more risk factors in the working environment from national sources. For the remaining 6 countries (GR,I,IRL,L,P,UK) data were only available through the Euro-survey.

- For the risk factors as presented in table 4.1.1a between 6 and 10 countries are able to provide data. In only some of these countries can data be differentiated by gender, age and economic sector, and to an even lesser degree by occupational categories.
- For the remaining risk factors data are only available from 5 or fewer countries.
- On only 5 output issues as presented in table 4.1.1b between 9 and 15 countries were able to provide data. As is the case with risk factors the output data can only be differentiated in some of these countries by gender, age and economic sector; and to an even lesser degree by occupational categories.
- For the remaining output indicators data are only available from 5 or fewer countries.

- The reliability of the available data is in general considered sufficient, i.e. concerning the total working population.
- As far as the working environment is concerned it should be underlined that practically all data are considered 'subjective', i.e. stemming from questionnaire-based surveys etc.. For the output data significantly more objective sources were found based on registrations.
- In general it should be noted that as far as output data is concerned, in many cases substantial under reporting can be assumed. This assumption can be derived both from 'experts' in the field as well as from remarks in the country reports. Therefore the reliability of output data, specifically occupational accidents and diseases is in all probability greatly affected by under reporting in several countries, however no overall assesment can be made as to the extent of this under reporting due to lack of specific information.

- The comparability of available data proves to be a major problem on most of the issues, due to a great variation in definitions and use of categories, with the exception of the risk factor 'noise' and to a lesser degree the output topics 'occupational accidents' and 'occupational diseases'. More in depth study might show however that some countries in the deviant group use definitions and categories similar to others in this group.
- Concerning the reference periods it can be concluded that in most countries the available data is fairly recent, especially with regard to output data.

5 *Figures on occupational health and safety across Europe*

In this chapter actual figures on occupational health and safety will be presented. As in chapter 4 the same criterion has been used in this chapter to present the data, i.e. if 6 or more countries have data available on a given item these data were taken up. This criterion has also been used in the more qualitative description of the aspects gender, age, occupation and economic sector.
Since the data from the various countries originate from different periods, the working population for each country in the year of reference of the main source has been taken up in Annex 7.

In the following tables of section 5.1, concerning physical, psychosocial and physiological exposures or risk factors, 'national data' and the data gathered in the framework of the 'First European Survey on the Work Environment' (also referred to as the 'Eurosurvey'; European Foundation, 1992) are presented and subsequently discussed.
In section 5.2 the available data on several health and safety output issues will be presented. Finally in section 5.3 an attempt has been made to draw some conclusions on the main risk factors and main risk groups across Europe. Furthermore some conclusions are drawn with regard to the data on health and safety output.

The figures in this chapter are presented without an extensive discussion on the quality and comparability of the data. This will be further evaluated in chapter 7. It is stressed beforehand however, that, in order to avoid false conclusions, cross-national comparison of the figures in the tables is only justified if the remarks belonging to the tables are thoroughly taken into account.

5.1 Working environment

During the screening of the available data from the national reports it became apparent that for the various risk factors a variety of conventions are currently in use. Consequently data on different aspects or sub-items belonging to the same risk factor were available for consolidation. In view of this fact a choice was made for presenting the sub-item with the highest frequency of exposed workers from each country which provided information.
Although from a methodological point of view a 'sound' approach, this has however reduced the possibility of comparisons between countries. In spite of the fact that time constraint made it impossible to go into further detail within the framework of this project, secondary analysis in the near future of the available data in this respect remains a distinct possibility.
The discussion of the data presented in the tables will therefore be merely descriptive. An emphasis will be put on the percentages of exposed workers to various aspects of a given risk factor and not so much on comparisons between countries.
Figures extracted from the Eurosurvey will be briefly compared to the 'national' data as far as the total average percentage of exposed workers is concerned. A particular problem in this respect is the fact that the total average percentage derived from the Eurosurvey refers to a different group of countries than the total averages of the respective risk factors. Nevertheless this comparison enables a cautious conclusion on the similarity between 'Euro' data and aggregated 'national' data for the various risk factors.

For the discussion of the variables, *gender, age, occupation* and *economic sector*, a limit was set in terms of a sufficient number of countries (6 or more) that had used a general or one single category for assessment of risk factors. It appeared that as a result from this condition only for the risk factors **noise** and **vibration** data on *gender* could be taken up.

Table 5.1.1: *Total number of workers and percentage of the working population exposed to physical risk factors in 16 European countries#*

Risk factors / Countries		Noise EF[1]	Noise This study	Vibrations This study	Climate EF[1]	Climate This study	Radiation This study
Austria (A)[3]	N %	- -	1,209,892 37.4%	312,779 9.7%	- -	≥ 723.501 ≥ 22.4%	- -
Belgium (B)[4]	N %	23.1%	≥ 183,156 ≥ 7.1%	110,585 4.3%	18.2%/26.7%	≥ 15,067 ≥ 0.6%	≥ 55,827 ≥ 2.2%
Denmark (DK)[5]	N %	25.7%	632,100 24.7%	≥ 145,900 ≥ 5.7%	14.5%/24.1%	≥ 775,000 ≥ 30.3%	[5] -
Spain (E)[6]	N %	31.5%	2,875,066 33.1%	660,136 7.6%	34.3%/42.8%	≥ 5,463,494 ≥ 62.9%	- -
Finland (FIN)[7]	N %	-	650,000 26.1%	≥ 230,000 ≥ 9.2%	- -	≥ 790,000 ≥ 31.8%	90,000 3.6%
France (F)[8]	%	31.4%	1,719,040 13.4%	455,040 3.6%	24.7%/38.1%	113,760 0.9%	214,880 1.7%
Germany (GER)[9]	N %	24.2%[2]	9,945,900 30%	- -	18.6%[2]/26.8%[2]	9,945,900 30%	- -
Greece	%	31.5%	-	-	42.8%/51.0%	-	-
Ireland (IRL)	%	31.3%	-	-	30.5%/36.2%	-	-
Italy (I)	%	25.3%	-	-	21.2%/31.9%	-	-
Luxembourg (L)	%	27.5%	-	-	16.0%/28.7%	-	-
Norway (N)[10]	N %	-	180,360 9%	120,240 6%	- -	≥ 681,360 ≥ 34%	- -
The Netherlands (NL)[11]	N %	19.6%	900,000 15.3%	615,000 10.5%	19.1%/29.4%	700,000 11.9%	30,000 0.1%
Portugal (P)	%	27.7%	-	-	32.8%/43.9%	-	-
Sweden (S)[12]	N %	-	808,656 20.4%	≥ 265,588 ≥ 6.7%	- -	≥ 1,185,236 ≥ 29.9%	166,488 4.2%
the United Kingdom (UK)	%	28.4%	-	-	22.3%/36.8%	-	-
Average % of exposed		**27.0%**	**≥ 23.9%**	**≥ 6.6%**	**22.7%/33.9%**	**≥ 26.4%**	**≥ 2.0%**

See introduction

Remarks:

General remark:
The average percentage exposed refers to a 'weighted' average percentage of the number exposed from countries that have provided data on that particular issue; the total number exposed were added up and divided by the total number of the working population of the countries that have provided data on that topic.

In some of the cells a '≥' sign has been added. This refers to the fact that for this item on more than one sub-item numbers of exposed persons were provided. For the purpose of presenting the data in this table, the sub-item with the highest frequency has been taken up. This means that *at least* this number of employees were exposed. E.g. if the concept 'climate' was utilised in a country in separate questions about exposure to heat and cold, the category with the highest frequency of exposed persons has been taken up.

European Foundation, 1992 (working population: employed and self-employed)

1) **Working Environment Survey (European Foundation, 1992):**
   ▸ the percentages on 'noise' refer to the category 'at least 25% of the time exposed to noise';
   ▸ the percentages on 'climate' refer to the categories 'at least 25% of the time exposed to bad weather' (first percentage) and 'at least 25% of the time exposed to heat or cold' (second percentage).

2)GER: the percentages of West-Germany (WD) and East-Germany (OD) were seperately calculated to the sample-size of each country. The number of respondents were added up and divided by the total sample-size (WD: 26,999 and OD: 8,531) of both countries (35,530 respondents).
   ▸ Noise: [(WD:24.2% x 26,999) + (OD:24.3% x 8,531)]/35,530 = 24.2%;
   ▸ Climate:
   a) bad weather: [(WD:17.6% x 26,999) + (OD: 21.8% x 8,531)]/35,531 = 18.6%;
   b) heat or cold: [(WD:24.3% x 26,999) + (OD: 34.9% x 8,531)]/35,531 = 26.8%.

This study:

3) A: the figures refer to the year 1985. The total working population is 3,234,500 (see table 5.1.1; ILO,1994). Only percentages according to gender were provided. These percentages were calculated to the Austrian male (1,957,400) and female (1,277,100) working population.
   Climate: the total number exposed refer to the category 'heat (inside)'.

4) B: the percentages are based on a total working population of 2,573,749 employees in 1991 (excluding 600,000 employees from the governmental and educational sector).
   Climate: the total number exposed refer to the category 'heat (inside)'.
   Radiation: the total number exposed refers to the category 'ultraviolet radiation'.

5) DK: the percentages are based on a total working population of 2,559,042 (1990). Concerning radiation no figures were included due to indicated unreliability.
   Vibrations: the total number exposed refers to the category 'whole body vibrations'.
   Climate: the total number exposed refers to the category 'dry air'.

6) E: the figures on noise and climate were extracted from a national working condition survey (1993). The percentages were extrapolated to a total working population (excl. the self-employed) of 8,686,000 (ILO, 1994b).
   Climate: the total number exposed refers to the category 'outdoor work: cold discomfort (winter)'.

7) FIN: the percentages are based on a total working population of 2,488,000 (ILO, 1994b). The numbers refer to unpublished data from a national Working Environment Survey of 1990. The data are estimated on the basis of an interview survey.
   Noise: the total number exposed refers to the category 'noise which prevents from hearing normal speech, at least ¼ of the working hours'.
   Vibration: the total number exposed refers to the category 'vibrations from manual machines at least ¼ of the working time'.

Climate: the total number exposed refers to the category 'draught, at least ¼ of the working time'.

8) F: the percentages are based on a working population of 12,640,000 (1986).
Noise: the total number exposed refers to the category 'noise of 85 dB or more'.
Vibrations: the total number exposed refers to the category 'vibrations from pneumatic instruments'.
Climate: the total number exposed refers to the category 'working in cold-storage chambers'.

9) GER: the percentages are based on a total working population of 33,153,000 (1991).

10) N: the figures were extracted from a National Survey of Working Conditions (1993), in which information about working conditions and working environment is collected from 3,682 respondents (only the employed). The percentages are extrapolated to a working population of 2,004,000 (ILO, 1994b).
Noise: the total number exposed refers to the category 'loud noise'.
Climate: the total number exposed refers to the category 'dry air'.

11) NL: the figures were extracted from a study made on basis on an inventory of available research material. The percentages are based on a total labour force of 5,885,000 workers (1990).
Radiation: the total number exposed refers to the category 'ionizing radiation'.

12) S: the figures were extracted from the Swedish Working Environment Study 1993. In this survey 11,732 respondents were interviewed. The percentages were extrapolated to a total working population of 3,964,000 (1993).
Noise: the total number exposed refers to the category 'noise which prevents from hearing normal speech, at least ¼ of the working hours'.
Vibrations: the total number exposed refers to the category 'whole body vibration'.
Climate: the total number exposed refers to the category 'dry air'.

Physical exposure

Table 5.1.1 shows that in ten countries (A,B,DK,E,F,FIN,GER,N,NL,S) there are data on exposure to noise, vibration, climate and radiation. The table also indicates that, when added up, exposure to noise is the most common physical risk factor in all ten countries: circa 19.1 million workers were exposed. Some 7.9 million workers have been exposed to various kinds of climatological hazards; some 3.3 million have been exposed to various forms of vibrations and some 740,000 workers to various forms of radiation.

With the limitations decribed in the introduction of 5.1. in mind with respect to the limited possibility of comparing countries the following observations on the risk factors can be made:

Noise:

Ten countries (A,B,DK,E,F,FIN,GER,N,NL,S) have provided data with respect to exposure to noise. Nine of these, with the exeption of Belgium, have used one single category or a general category to measure exposure to noise.

The average percentage of workers exposed to various aspects of the risk factor noise, extracted from this study, amounts at least to circa 24%. Although rather roughly calculated this percentage closely resembles the average percentage of the European Working Environment Survey (27%).

The range of percentages on exposure to various aspects of noise-hazards is broad and varies from 7% to 37% of the national working populations. Relatively high percentages are found in Austria, Denmark and Germany (on average 30% or more), while in Belgium and Norway rather low percentages exposed to any kind of noise-hazard were found (below 10%).

Gender:
Six countries (A,DK,FIN,N,NL,S) have provided data on exposure to noise according to gender. In Austria and the Netherlands the percentage of exposed male workers (29%) is significantly higher than the percentage of female workers (9%) who were exposed to noise. In Finland and Sweden the percentages of exposed male workers are respectively 35% and 28% versus respectively 18% and 13% for the number of exposed female workers.

Vibration:
Nine countries (A,B,DK,E,F,FIN,N,NL,S) have provided data on exposure to vibration. Six of these countries have used a single category or a general category concerning exposure to vibrations. Of the nine countries some 6.6% or more of the working population have been exposed in general or specifically to some kind of vibration during their work. The range varies from 4.3% up to 10.5% exposed workers. The Netherlands, Austria and Finland are the countries with the highest percentages (more than 9%) of exposed workers.
As this aspect was not included in the Eurosurvey no average percentage of exposed workers from this survey was available.

Gender:
Six countries (A,DK,FIN,N,NL,S) have provided data on exposure to vibration according to gender. In Finland the difference of exposure between male and female workers to vibration of manual machines is 13% (male: 16% versus female: 3%). In Austria the difference is 9.7% (male: 14% versus female: 4%) and in Sweden 11% (male: 12% versus female: 1%). Finally, the difference between exposed male and female workers in Denmark and the Netherlands is 9% (DK: male: 10% versus female 1%; NL: male: 11% versus female: 2%).

Climate:
Ten countries (A,B,DK,E,F,FIN,GER,N,NL,S) provided data on exposure to climate as a physical risk factor. Only three countries used a general or single category to measure exposure to climatological risk factors. The range of percentages on exposure to climate varies from 0.6% to 62.9%. Spain reports a relatively high percentage of exposure (circa 63%). Relatively low percentages of exposed numbers were found in Belgium and France (less than 1%).
Finally, the average percentage of workers exposed to climate, extracted from this study, amounts to at least (26%), which is more or less consistent with the average percentages of the European Working Environment Survey (22.7%/33.9%).

Radiation:
Five countries (B,F,FIN,NL,S) have provided data on this physical risk factor. Only in Belgium more exposure categories were used. The other countries use a general or one specific category to measure exposure to radiation. The range of percentage of exposure varies from 0.1% to 4.2%. The average percentage over these 5 countries of exposed workers to various kinds of radiation is at least 2%.
Comparable data from the Eurosurvey were not available.

Table 5.1.2: ***Total number of workers and percentage of the working population exposed to psychosocial risk factors in 16 European countries#***

Risk factors / Countries		Working pace		Job content		Working hours		Influence and control over own work		Social interaction	
		EF[1]	This study	EF[1]	This study	EF[1]	This study	EF[1]	This study	EF[1]	This study
Austria (A)[3]	N %	-	1,355,256 41.9%	-	326,685 10.1%	-	485,175 15.0%	-	- -	-	889,191 27.8%
Belgium (B)[4]	N %	43.8%/34.3%	- -	37.3%	- -	17.9%/14.2%	[4] [4]	34.6%/33.4%	- -	14.4%	- -
Denmark (DK)[5]	N %	59.7%/61.6%	≥ 1,520,000 ≥ 59.4%	40.4%	685,800 26.8%	14.0%/9.8%	≥ 409,000 ≥ 16.0%	34.4%/29.4%	478,500 18.7%	13.3%	≥ 954,500 ≥ 37.3%
Spain (E)[6]	N %	37.0%/35.1%	≥ 2,857,694 ≥ 32.9%	52.1%	≥ 4,073,734 ≥ 46.9%	27.0%/15.3%	≥ 1,537,422 ≥ 17.7%	44.4%/38.4%	≥ 5,819,620 ≥ 67.0%	14.1%	≥ 2,788,206 ≥ 23.1%
Finland (FIN)[7]	N %	-	- -	-	≥ 1,000,000 ≥ 40.2%	-	- -	-	≥ 1,244,000 ≥ 50%	-	≥ 895,680 ≥ 36%
France (F)	%	38.9%/42.1%	-	43.1%	-	20.9%/17.9%	-	35.5%/36.6%	-	24.3%	-
Germany (GER)[8]	N %	56.1%[2]/61.6%[2]	18,234,150 55%	59.3%[2]	≥ 15,913,440 ≥ 48%	17.6%[2]/15.6%[2]	≥ 6,299,070 ≥ 19%	43.9%[2]/44.7%[2]	11,603,550 35%	12.2%	4,972,950 15%
Greece (GR)	%	68.0%/62.9%	-	60.1%	-	41.9%/24.7%	-	43.9%/40.1%	-	26.6%	-
Ireland (IRL)	%	38.5%/48.8%	-	50.4%	-	34.0%/32.6%	-	33.4%/27.1%	-	10.5%	-
Italy (I)	%	43.2%/30.2%	-	42.3%	-	29.2%/11.0%	-	36.9%/28.6%	-	28.6%	-
Luxembourg (L)	%	35.6%/39.4%	-	38.0%	-	15.7%/15.8%	-	41.3%/38.8%	-	12.6%	-
Norway (N)[9]	N %	-	≥ 1,342,680 ≥ 67%	-	≥ 1,643,280 ≥ 82%	-	≥ 801,600 ≥ 40%	-	≥ 1,022,040 ≥ 51%	-	≥ 841,680 ≥ 42%
The Netherlands (NL)[10]	N %	56.8%/37.4%	3,245,000 55%	51.1%	≥ 2,242,000 ≥ 38%	12.7%/13.3%	≥ 3,199,500 ≥ 54%	25.7%/22.0%	1,182,885 20.1%	8.5%	- -
Portugal (P)	%	55.7%/33.6%	-	52.2%	-	29.5%/15.9%	-	41.9%/35.3%	-	21.3%	-
Sweden (S)[11]	N %	-	- -	-	≥ 2,009,748 ≥ 50.7%	-	≥ 808,656 ≥ 20.4%	-	≥ 1,692,628 ≥ 42.7%	-	≥ 2,227,768 ≥ 56.2%
United Kingdom(UK)	%	42.3%/64.0%	-	53.7%	-	22.4%/27.6%	-	30.4%/28.5%	-	16.6%	-
Average % of exposed		**47.3%/48.9%**	**≥ 51.4%**	**50.9%**	**≥ 45.0%**	**14.8%/17.6%**	**≥ 29.4%**	**37.6%/35.2%**	**≥ 39.2%**	**18.2%**	[12]

See introduction

Remarks:

General remark:
The average percentage exposed refers to a 'weighted' average percentage exposed of countries that have provided data on the topic; the total number of exposed were added up and divided by the total number of the working population of the countries that have provided data on that topic.

In some of the cells a '≥' sign has been added. This refers to the fact that for this item on more than one sub-item numbers of exposed persons were obtained. For the purpose of presenting this data the sub-item with the highest frequency has been taken up. This means that *at least* this number of employees were exposed. E.g. if the concept 'working pace' was utilised in a country in separate questions about exposure to repetitive work and working at high speed, the category with the highest frequency of exposed persons has been taken up.

European Foundation, 1992 (working population: employed and self-employed):

1) **Working Environment Survey (European Foundation, 1992):**

- **working pace:** the first percentage refers to the category 'working at least 25% of the time at high speed'. The second percentage refers to the category 'working at least 25% of the time to tight deadlines';
- **job content:** the percentage refers to the category 'doing at least 25% of the short time repetitive tasks';
- **working hours:** the first percentage refers to the category '% of employed and self-employed workers who work more than 45 hours per week'. The second percentage refers to the category 'working at least 25% of the time at night'.
- **influence and control over own work:** the first percentage refers to the category '% of workers who aren't able to change tasks/work method'. The second percentage refers to the category '% of workers who aren't able to change speed/rate of work'.
- **social interaction:**the percentage refers to the category 'do not have sufficient support from superiors or colleagues to carry out my own work'.

2)GER: the percentages of West-Germany (WD) and East-Germany (OD) were seperately calculated to the sample-size of each country. The number of respondents were added up and divided by the total sample-size (WD: 26,999 and OD: 8,531) of both countries (35,530 respondents).

- working pace:
  a) high speed: (WD:55.5% x 26,999) + (OD:57.9% x 8,531)/35,530 = 56.1%;
  b) tight deadlines: (WD:59.7% x 26,999) + (OD:63.6% x 8,531)/35,530 = 61.6%;
- job content: (WD:58.1% x 26,999) + (OD:63.1% x 8,531)/35,530 = 59.3%;
- working hours:
  a) > 45 hours/week: (WD:12.2% x 26,999) + (OD:9.3% x 8,531)/35,530 = 11.5%;
  b) working at night: (WD:15.1% x 26,999) + (OD:17.2% x 8,531)/35,530 = 15.6%;
- influence and control over own work:
  a) change tasks: (WD:43.4% x 26,999) + (OD:45.3% x 8,531)/35,530 = 43.9%;
  b) change speed: (WD:44.4% x 26,999) + (OD:45.8% x 8,531)/35,530 = 44.7%.
- social interaction: (WD:10.5% x 26,999) + (OD:17.7% x 8,531)/35,530 = 12.2%;

This study:

3) A: the figures refer to the year 1985. The total working population is 3,234,500 (see table 5.1.1; ILO, 1994b). Only percentages according to gender were provided. These percentages were calculated to the Austrian male (1,957,400) and female (1,277,100) working population.
Working pace: the number of exposed refers to the category 'working under pressure of time'.

Job content: the number exposed rèfers to the category 'monotonous work'.
Working hours: the number exposed refers to the category 'regular overtime'.
Social interaction: the number exposed refers to the category 'a lot of contacts with clients'.

4) B: Concerning 'working hours' data have been taken up in the national report. There are data on gender, age and economic sectors regarding the following six categories: (1) working in shifts; (2) working in the evening; (3) working at night; (4) working on saturdays; (5) working on sundays and (6) working at home. No total number of workers was explicitely taken up in the national report. Therefore the number of workers could not be taken up in the table above.

5) DK: the percentages are based on a total work force of 2,559,042 (1990).
Working pace: the number exposed refers to the category 'job demands full concentration for more than ¾ of the working time'.
Job content: the number exposed refers to the category 'no opportunity to learn and qualify oneself'.
Working hours: the number exposed refers to the four categories 'shiftwork, irregular working hours and night work and evening work', which have been added up.
Social interaction: the number exposed refers to the category 'communication with work mates impossible during working hours for at least ¼ of the working time'.

6) E: the figures on form of payment, job content and working hours were extracted from a national working condition survey (1993). The percentages were extrapolated to a total working population (excl. the self-employed) of 8,686,000 (ILO, 1994b).
Working pace: the number exposed refers to the category 'at least 50% of the time doing short repetitive tasks'.
Job content: the number exposed refers to the category 'unskilled work'.
Working hours: the number exposed refers to the four categories 'shiftwork, double shifts, triple shifts, nightwork'), which have been added up.
Influence/control: the number exposed refers to the category 'insufficient participation'.
Social interaction: the number exposed refers to the category 'difficult communication with co-workers due to isolated work'.

7) FIN: the numbers refer to unplubished data from an national Working Environment Survey of 1990. The data are estimated on the basis of an interview survey. The percentages are based on a total working population of 2,488,000 (ILO, 1994b).
Job content: the number exposed refers to the category 'repetitive work'.
Influence and control: the number exposed refers to the category 'can't influence on work tasks'.Social interaction: the number exposed refers to the category 'lack of support by colleagues'.

8)GER: the percentages are based on a working population of 33,153,000.
Job content: the number exposed refers to the category 'repetitive work'.
Working hours: the number exposed refers to the category 'shiftwork'.
Social interaction: the number of exposed refers to the category 'seldom team work'.

9) N: the figures were extracted from a National Survey of Working Conditions (1993), in which information about working conditions and working environment is collected from 3,682 respondents (only the employed). The percentages are extrapolated to a total working population of 2,004,000 workers in 1990 (ILO, 1994b).
Working pace: the number exposed refers to the category 'percentage of employees who for half of their working time or more are keeping a workingplace totally controlled by deadlines and fixed routines'.
Job content: the number exposed refers to the category 'percentage of employees who for half of their working time or more must be aware to prevent errors or accidents'.
Working hours: the number exposed refers to the three categories 'permanent evening/nightwork, shiftwork, regulated flexible hours', which have been added up.
Influence and control over own work: the number exposed refers to the category 'percentage who to a large extent cannot plan their daily work themselves'.
Social interaction: the number exposed refers to the category 'percentage who often or now and then experience conflicts or poor relations between management and employees'.

10)NL: the percentages on 'working pace are based on a total working population of 5,900,000 workers (1992), the percentages on 'job content' are based on a total working population of 5,885,000 (1990) and the percentages on 'working hours' are based on a total working population (in 1993) of 5,925,000 (ILO, 1994b).
Working pace: the number exposed refers to the category 'working at high speed'.
Jobcontent: the number exposed refers to the category 'monotonous work'.
Working hours: the number exposed refers to two categories 'deviant working hours' and shift work', which have been added up.

11)S: the figures were extracted from the Swedish Working Environment Study 1993. In this survey 11,732 respondents were interviewed. The percentages were extrapolated to a total working population of 3,964,000 (1993).
Job content: the number exposed refers to the category 'far to much to do'.
Working hours: the number exposed refers to the five categories 'working evenings between 6 and 1pm, (2) nights between 1pm and 6am, (3) double shifts/triple shifts, (4) rotating shifts and (5) other hours', which have been added up.
Influence and control over own work: the number exposed refers to the category 'can't usually not partly decide on their own when various tasks are to be done'.
Social interaction: the number exposed refers to the category 'are occasionally forced to cope on your own without being able to count on the help of others around you in critical situations (roughly ¼ of the time or more)'.

12) Due to the extensive diversity of both positive and negative aspects mentioned in relation to this topic and the limited number of remaining countries, no average percentage of exposed persons has been calculated.

Psychosocial exposure

Table 5.1.2 shows that in eight countries (A,DK,E,FIN,GER,N,NL,S) there are data (from the national reports) on exposure to one or more of the following psychosocial risk factors: working pace, job content, working hours, influence and control over own work and social interaction.
The table also indicates the highest total number of exposures is for working pace: more than 29 million workers in the six countries with data. Some 20 million workers have problems regarding job content, more than 19 million regarding influence and control over own work, more than 10 million with respect to social interaction and finally over 10 million workers indicated problems regarding working hours.

With the limitations decribed in the introduction of 5.1. in mind with respect to the limited possibility of comparing countries the following observations on the risk factors can be made:

Working pace:

Six countries (A,DK,E,GER,N,NL) provided data on exposure to excessive working pace. Three countries (DK,E,N) have used more than one category to assess this risk factor. The other countries have used a general or one specific category to measure this.
The range of percentages of exposed numbers of workers to working pace varies from 33% up to 67%. Relatively high percentages were found in Norway and Denmark (on average 59% or more).
The average percentage of workers exposed to working pace over the six countries is slightly higher (54.4%) than the average percentages that were extracted from the European Working Environment Survey (47.3%/48.9%).

Job content:
Eight countries (A,DK,E,FIN,GER,N,NL,S) have provided data (in their national reports) on job content as a psychosocial risk factor. Only two countries (A,DK) use one single or a general category to measure this risk factor. The percentage of total number of exposed workers (within each country) to job content varies from 10% up to 82%. Especially Norway shows a high percentage (82%), however the category used in Norway in this respect is of a different nature than the categories used in the other countries (see remarks). The other countries, especially Spain, Germany and Sweden, show relatively high percentages (over 46%).
Finally, the average percentage of workers exposed to risks within the context of 'job content' is slightly lower (45%) than the average percentage that was extracted from the European Working Environment Survey (50.9%).

Working hours:
Eight countries (A,B,DK,E,GER,N,NL,S) have provided data on working hours. Only Austria uses a single category for measurement of this risk factor. The percentage of total number exposed of all workers (within each country) to working hours as a psychosocial risk factor varies from 15% up to at least 54%.
Especially in Norway and the Netherlands relatively high percentages (over 40%) of workers were found to be exposed to some aspect of working hours as a risk factor.
Finally, the average percentage of workers over all seven countries exposed to risks within the framework of working hours is over 10% higher (29.4%) than the average percentages that were extracted from the European Working Environment Survey (14.8%/17.6%).

Influence and control over own work:
Table 5.1.2 shows that seven countries (DK,E,FIN,GER,N,NL,S) provided data on this risk factor. Three countries (DK,GER,NL) have used a general or one single category to measure exposure to this risk factor. The range of percentages of exposed workers to various aspects of this psychosocial risk factor varies from 18.7% up to 67%. Relatively high percentages are reported from Finland, Norway and Spain (over 50%).
Finally, the average percentage of workers over these seven countries exposed to this risk factor (39.2%) is more or less consistent with the average percentages that were extracted from the European Working Environment Survey (37.6%/35.2%).

Social interaction:
Although seven countries provided data in this respect, the data from both Austria and Germany were ambiguous. Consequently this leaves only five countries with more or less comparable data. These five countries all used more than one category in assessing this risk factor.
The range of percentages of exposure to social interaction (as a psychosocial risk factor) varies from 15% up to 56%. Especially Sweden reported a high percentage of exposed numbers (at least 56%) to one aspect of the risk factor social interaction.
Finally, the average percentage of workers exposed to this risk factor was not calculated due to the limited numbers and the apparent ambiguity of the available data.

Table 5.1.3: ***Total number of workers and percentage of the working population exposed to physiological risk factors in 16 European countries#***

Risk factors / Countries		PHYSIOLOGICAL EXPOSURE			
		Working posture and movements		Manual lifting,handling, pushing and pulling	
		EF[1]	This study	EF[1]	This study
Austria (A)[3]	N		-		≥ 847,798
	%	-	-	-	≥ 25.5%
Belgium (B)	N		-		-
	%	34.6%	-	25.7%	-
Denmark (DK)[4]	N		≥ 1,389,500		404,300
	%	35.0%	≥ 54.3%	28.9%	15.8%
Spain (E)[5]	N		≥ 3,465,714		1,172,610
	%	43.1%	≥ 39.9%	32.8%	13.5%
Finland (FIN)[6]	N		≥ 520,000		470,000
	%	-	≥ 20.9%	-	18.9%
France (F)	N		-		-
	%	46.3%	-	37.9%	-
Germany (GER)[7]	N		8,288,250		9,614,370
	%	47.0%[2]	25%	28.4%[2]	29.0%
Greece (GR)	N		-		-
	%	68.9%	-	42.8%	-
Ireland (IRL)	N		-		-
	%	39.1%	-	35.6%	-
Italy (I)	N		-		-
	%	42.1%	-	25.2%	-
Luxembourg (L)	N		-		-
	%	27.2%	-	22.1%	-
Norway (N)[8]	N		≥ 1,102,200		≥ 320,640
	%	-	≥ 55%	-	≥ 16%
The Netherlands (NL)[9]	N		≥ 2,707,100		1,652,000
	%	22.1%	≥ 46%	21.6%	28%
Portugal (P)	N		-		-
	%	55.8%	-	32.0%	-
Sweden (S)[10]	N	-	≥ 1,089,028		705,592
	%	-	≥ 27.7%	-	17.8%
the United Kingdom (UK)	N		-		-
	%	31.9%	-	32.0%	-
Average % of exposed		**42.2%**	**≥ 31.6%**	**30.7%**	**≥ 24.5%**

See introduction

Remarks:

General remark:
The average percentage of exposed refers to a 'weighted' average percentage of the exposed of countries that have provided data on the issue; the total number exposed were add up and divided by the total number of the working population of the countries that have provided data on that issue.

In some of the cells a '≥' sign has been added. This refers to the fact that for this item on more than one sub-item numbers of exposed persons were obtained. For the purpose of presenting this data the sub-item with the highest frequency has been taken up. This means that *at least* this number of employees were exposed. E.g. if the concept 'working posture and movements' was utilised in a country in separate questions about exposure to working in a standing position and working in a sitting position, the category with the highest frequency of exposed persons has been taken up.

European Foundation, 1992 (working population: employed and self-employed):

1) **Working Environment Survey (European Foundation, 1992):**
▸ the percentages on 'working postures and movements' refer to the category 'at least 25% of the time working in painful positions';
▸ the percentages on 'manual lifting, handling, pushing and pulling' refer to the categorie 'at least 25% of the time carrying heavy loads'.

2)GER: the percentages of West-Germany (WD) and East-Germany (OD) were seperately calculated to the sample-size of each country. The number of respondents were added up and divided by the total sample-size (WD: 26,999 and OD: 8,531) of both countries (35,530 respondents).
▸ Working postures: [(WD:46.6% x 26,999) + (OD:48.4% x 8,531)]/35,530 = 47.0%;
▸ Manual handling: [(WD:26.7% x 26,999) + (OD: 33.8% x 8,531)]/35,531 = 28.4%.

This study:

3) A: the figures refer to the year 1985. The total working population is 3,234,500 (see table 5.1.1; ILO,1994). Only percentages according to gender were provided. These percentages were calculated to the Austrian male (1,957,400) and female (1,277,100) working population.
Manual handling: the number exposed refers of the category 'other heavy physiological load'.

4) DK: the percentages are based on a total working population of 2,559,040 (1990).
Working postures and movements: the number exposed refers to the category 'standing more than ¼ of the working time'.
Manual handling: the number exposed refers to the category 'lifting burdens of > 20 kg more than seldom'.

5) E: the figures on working postures/movements and manual lifting, handling, pushing and pulling were extracted from a national working condition survey (1993). The percentages were extrapolated to a total labour force (excl. the self-employed) in 1993 of 8,686,000 (ILO, 1994b).
Working postures/movements: the number exposed refers to the category 'standing and walking'.
Manual handling: the number exposed refers to the category 'heavy work'.

6) FIN: the numbers refer to unplubished data from an national Working Environment Survey of 1990. The data are estimated on the basis of an interview survey. The percentages are based on a total working population IN 1990 of 2,488,000 (ILO, 1994b).
Working postures/movements: the number exposed refers to the category 'repetitive, monotonous work movements'.
Manual handling: the number exposed refers to the category 'heavy lifting'.

7)GER: the percentages are based on a working population in 1991 of 33,153,000 (ILO, 1994b).
Working postures/movements: the number exposed refers to the category 'working in aggravating postures'.
Manual handling: the number exposed refers to the category 'manual handling of loads that weight more than 20 kg'.

8) N: the figures were extracted from a National Survey of Working Conditions (1993), in which

information about working conditions and working environment is collected from 3,682 respondents (only the employed). The percentages are extrapolated to a total working population of 2,004,000 workers in 1990 (ILO, 1994b).

Working postures/movements: the number exposed refers to the category 'working in a sitting position'.

Manual handling: the number exposed refers to the category 'at least 5 times a day lifting something that weighs at least more than 20 kg'.

9)NL: the figures on the category 'manual lifting, handling,etc.' were extracted from a study on the life situation of the Dutch population (1992). The percentages have been extrapolated to a total labour force of 5,900,000 workers. The figures on 'working postures and movements' have been extrapolated to a total working population in 1990 of 5,885,000 (ILO, 1994b)

Working postures/movements: the number exposed refers to the category 'working in a sitting position'.

Manual handling: the number exposed refers to the category 'physical hard labour'.

10)S: the figures were extracted from the Swedish Working Environment Study 1993. In this survey 11,732 respondents were interviewed. The percentages were extrapolated to a total working population of 3,964,000 (1993).

Working postures/movements: the number exposed refers to the category 'roughly ¼ of the time or more work bending forward without supporting yourself with your hands or arms'.

Manual handling: the number exposed refers to the category 'have to lift at least 20 kg several times a day'.

Physiological exposure

Table 5.1.3 shows that in eight countries (A,DK,E,FIN,GER,N,NL,S) there are data on exposure to working postures/movements and/or manual lifting, handling, pushing and pulling as a physiological risk factor. The table also indicates that, when added up, exposure to working postures/movements is the most common physiological risk factor among these countries: circa 16 million workers. Some 15 million workers have been exposed to manual lifting, handling, etc.

With the limitations decribed in the introduction of 5.1. in mind with respect to the limited possibility of comparing countries the following observations on the risk factors can be made:

Working postures and movements:

Seven countries (DK,E,FIN,GER,N,NL,S) have provided data on exposure to this risk factor. Only in Germany a single category was used for measurement. The range of percentages on exposure varies from 21% to 55%. Especially Norway and Denmark reported high percentages (at least over 54%) of exposed workers to various kinds of aspects of this risk factor.

Finally, the average percentage of workers exposed to this risk factor (31.6) is circa 10% lower than the average percentage that was extracted from the European Working Environment Survey (42.2%).

Manual lifting, handling, pushing and pulling:

Eight countries (A,DK,E,FIN,GER,N,NL,S) have provided data. Only in Austria and Norway has more than one category been used to assess this risk factor. The range of percentages on exposure to manual lifting varies from 14% to 29%. Relatively high percentages of exposed workers are found in the Netherlands and Germany (over 28%).

Finally, the average percentage of workers exposed to this risk factor is slightly lower (24.5%) than the average percentage that was extracted from the European Working Environment Survey (30.7%).

5.2 Health and safety output

In this section the various aspects on which sufficient data has been provided, will be presented. As far as the variables *gender, age, occupation* and *economic sector* are concerned, it should be noted that due to an overall lack of information only a limited number of descriptions on these variables have been included. *Occupation* therefore has not been included at all and *economic sector* is included on only one output indicator. Furthermore if information was available, especially in case of the variable *age*, the categories used in the various countries were rather diverse. Consequently no descriptions of this variable in relation to the various output indicators have been included.

Table 5.2.1a: ***Number of occupational accidents in 1 year and rate of people (per 10,000 workers) that have had an occupational accidents in 15 European countries#***

Countries	Item	OCCUPATIONAL ACCIDENTS		
		All accidents	**Fatal accidents**	**With work interruption**
Austria (A)	N	229,712	335	-
	rate	637[1]	0.9[1]	-
Belgium (B)	N	230,925	271	157,329
	rate	897[2]	1.1[2]	611[2]
Denmark (DK)	N	-	61	44,247
	rate	-	0.2[3]	170[3]
Spain (E)	N	1,116,552	1239	628,640
	rate	903[4]	1.0[4]	508[4]
Finland (FIN)	N	103,533	64	69,225
	rate	471[5]	0.3[5]	315[5]
France (F)	N	813,080	1,024	750,058
	rate	564[6]	0.7[6]	519[6]
Germany (GER)	N	1,715,400	2,227	1.469,624
	rate	464	0.6[7]	398[7]
Greece (GR)	N	28,900	203	-
	rate	80[8]	0.6[8]	-
Italy (I)	N	-	1,937	972,468
	rate	-	0.9[9]	450[9]
Ireland (IRL)	N	(3,606)	(64)	(3542)
	rate	(32)[10]	(0.6)[10]	(32)[10]
Luxembourg (L)	N	19,729	23	-
	rate	1,002[11]	1.2[11]	-
Netherlands (NL)	N	-	59	65,600
	rate	-	0.1[12]	107[12]
Portugal (P)	N	243,616	168	243,448
	rate	547[13]	0.4[13]	546[13]
Sweden (S)	N	-	103	38,484
	rate	-	0.3[14]	97[14]
United Kingdom (UK)	N	168,813	430	140,365
	rate	66[15]	0.2[15]	55[15]
Range of total	N	**19,729(L) - 1,715,400(GER)**	**23(L) - 2,227(GER)**	**38,484(S) - 1,429,624(GER)**
	rate	**66(UK) - 1,002(L)**	**0.1(NL) - 1.4(E)**	**55(UK) - 724(E)**

See introduction

Remarks:

General remarks:
In 6 countries (B,DK,E,F,NL,S) occupational accidents are registered only if the injured person is absent from work at least one day after the accident. In 5 countries (A,FIN,GER,I,UK) the figures relate to three days absence and more. For 4 countries (GR,IRL,L,P) occupational accidents are not defined. Some countries moreover include commuting accidents (see below).
Furthermore the figures are based on a total working population (i.e. the employed and self-employed), except Belgium (600,000 employees are excluded) and Spain (due to the fact that the data were extracted from a survey that was done by only employed workers, the self-emloyed were excluded).
Finally there is reason to believe that in several countries there is a significant underreporting. This was especially indicated for Ireland, the Netherlands and the UK. These figures are therefore not reliable.

N = the number of occupational accidents.
Rate = ⟦the number of accidents : the total working population⟧ x 10,000.

All accidents, fatal accidents and accidents with work interruption:

1) A: the number of accidents (229,712) relate to the reference period of 1993. For fixing the rate a total working population of 3,608,100 people has been used (1993).
2) B: the figures refer to accidents with temporary work interruption. The category 'accidents with permanent work interruption' is not taken up into this table.
The number of accidents relate to the reference period of 1993. Total working population: 2,532,719 (1993). Regarding fatal accidents and accidents with work interruption data are available also according to economic sector.
3) DK: the number of accidents relate to the reference period of 1992. Total working population: 2,609,859 (1992).
4) E: the number of accidents relate to the reference period of 1992. Total working population: 12,366,000 (1992).
5) FIN: the number of accidents relate to the reference period of 1992. Total working population: 2,199,000 (1992).
6) F: the number of accidents relate to workers in 15 large sectors with a total working population of 14,440,400 employees (1992).
7) GER the number of accidents refer to accidents at the working place. In 1993 45,342 claims (out of 1.7 million accidents) were compensated. 80% of these compensations relate to accidents at the working place and 20% to commuting accidents. Total working population: 36,940,000 (1992).
8) GR: the number of accidents relate to the reference period of 1991. Total working population: 3,632,400.
9) I: employees and the self-employed are insured against accidents at work. Non-manual workers in general are not covered unless they are working with certain electrical equipment. Also railroad workers, offshore workers and flight staff of airlines are insured. Commuting accidents are generally not included in the figures. However, road accidents occuring in the course of work are included. The number of accidents relate to the reference period of 1991. Total working population is 21,595,000 (1991).
10) IRL there is reason to believe that there is a significant under-reporting of accidents. That is the reason the figures are placed between brackets. This applies also for the category 'causes of occupational accidents'. Information from the Health and Safety Authority (HSA) indicates that a third or less of all accidents are reported to HSA. Reference period: 1992. Total working population: 1,125,800 (1992).
11) L: the figures relate to the reference period of 1993. There are no data avalaible on the total working population of that year. That's why the number of occupational accidents are based on a total working population of 196,800 (1992).
12) NL: the figures are based on a total working population of 6,155,000 (1989). Underreporting is indicated.
13) P: the figures are based on a total working population of 4,457,600 (1993).

14) S: the working population covers all persons who are either gainfully employed or unemployed. Conscripted personnel (Armed forces) are not included for the categories 'all accidents' and 'causes of accidents'. Total working population of 3,964,800 (1993).

15)UK: there is reason to believe there is a significant under-reporting of accidents. The figures are based on a total working population of 25,629,100 (1992).

Occupational accidents

Table 5.2.1a shows that fifteen countries (A,B,DK,E,F,FIN,GER,GR,I,IRL,L,NL,P,S,UK) have provided data on one or more aspects of occupational accidents (all accidents, fatal accidents, accidents 'with work interruption').

All accidents:

Ten countries have provided data (A,B,E,F,FIN,GR,IRL,L,P,UK) regarding the category 'all occupational accidents'. The range of the rates varies extraordinarily from 32 in Ireland up to 1002 in Luxembourg per 10,000 workers. It should however be noted that for Ireland substantial underreporting was indicated. Relatively high rates are furthermore reported in Belgium and Spain, on average a rate of 900.

Fatal accidents:

Fifteen countrees (A,B,DK,E,F,FIN,GER,GR,I,IRL,L,NL,P,S,UK) have provided data regarding fatal accidents. The range of the rates varies from 1 to 12 fatal accident per 100,000 (employed + self-employed) workers. Relatively high rates are reported in Luxembourg and Belgium, respectively 12 and 11 fatal accidents per 100,000 workers.

Gender:

Eight countries (B,DK,E,GER,I,NL,S,UK) have provided data on fatal accidents according to gender. In all countries the greatest part of all fatal accidents occur within the male working population. In five countries (DK,E,GER,I) the rate for fatal accidents for the male workers is substantially higher than of the female working population. In Denmark and Germany the male rate is 5 times higher than the female rate; in Italy over 7 times and in Spain over 15 times higher than the rates for females.

Economic sector:

Seven countries (B,DK,E,GER,I,IRL,S) indicated that data according to economic sector were available. However only five countries (DK,E,I,IRL,S) actually provided information in their national reports in time.

Concerning the sector *'agriculture, forestry and fishing'* especially the rates of fatal accidents in Italy, Ireland and Spain appear to be relatively high, respectively 21, 17 and 10 fatal accidents per 100,000 workers.

Within the sector *'energy and water'* especially the rates of fatal accidents in Spain (30), Ireland (22) and Denmark (22) appear to be relatively high.

The rates of fatal accidents within the sector *'extraction and processing of non-energy minerals, chemical industries'* appear to be relatively high in Italy and Spain, respectively 22 and 10 fatal accidents per 100,000 workers. For the economic sector *'metal manufacturing, mechanical and electrical industry'* especially the rates of fatal accidents appear to be relatively high in Spain and Ireland; in both countries 10 fatal accidents per 100,000 workers. Concerning the sector *'building and civil engineering'* especially the rates of Spain, Italy and Ireland appear to be relatively high, respectively 20, 19 and 14 fatal accidents per 100,000 workers. Finally, within the economic sector *'transport and communication'* especially the rates in Spain, Italy and Ireland appear to be relatively high, respectively 20, 15 and 11 fatal accidents per 100,000 workers.

The remaining countries in these six economic sectors reported less than 10 fatal accidents per 100,000 workers. Concerning the other four economic sectors all five countries reported less than 10 fatal accidents per 100,000 workers.

Accidents with work interruption:
Table 5.2.1 shows that twelve countries (B,DK,E,F,FIN,GER,I,IRL,NL,P,S,UK) have provided data regarding accidents with work interruption. The rate varies from 55 in the UK up to 611 accidents with work interruption per 10,000 workers in Belgium. However as far as the UK is concerned significant underreporting is indicated (as is the case in Ireland). Relatively high rates are further reported in Spain, France and Portugal (over 500).

Gender:
Nine countries (B,DK,E,F,FIN,FER,I,NL,S) have provided data according to gender. In all nine countries the percentage of male workers that has had an accident 'with work interruption' is higher than the percentage of female workers. The range of rates for male/female workers varies respectively from 130 to 70 per 10,000 workers in Sweden up to 880 versus 190 in Belgium. Other countries with relatively high differences between male and female workers are Spain and Germany, respectively 660 versus 190 and 570 versus 160.

Age:
Eight countries (B,DK,E,F,FIN,GER,I,NL) have provided data according to age-categories. However the age categories in these countries differ too much (none provided age categories corresponding to those of the matrix), therefore it is not feasible to make sensible comparisons.

Economic sectors:
Seven countries (B,DK,E,GER,I,NL,S) reported that there were data on this issue. To describe the accidents with work interruption according to economic sectors the three economic sectors for each country with the highest percentages of accidents with work interruption are listed below. The rates of accidents with work interruption are given between brackets.

Belgium:	data were not provided in time.
Denmark:	(1) Manufacturing (3.5%); (2) Energy and water (3.0%) and (3) Transport and communication (2.4%);
Spain:	(1) Energy and water (12.9%); (2) Metal manufacturing (11.8%) and (3) Building and civil engineering (8.8%).
Germany:	(1) Building and civil engineering (18.2%); (2) Banking (11.8%) and (3) Metal manufacturing (8.8%);
Italy:	(1) Extraction and processing of non-energy producing minerals (13.6%); (2) Agriculture (12.4%) and (3) Building and civil engineering (7.5%);
The Netherlands:	(1) Building and civil engineering (4.2%); (2) Metal manufacturing, mechanical and electrical industry (2.5%) and (3) Other manufacturing industries (1.7%);
Sweden:	(1) Metal manufacturing, mechanical and electrical industry (4.0%); (2) Extraction and processing of non-energy producing minerals (2.3%) and (3) Building and civil engineering (1.7%).

Table 5.2.1b: ***Overview on 'causes' of occupational accidents in 1 year in 11 (out of 16) European countries***

Countries / Item	Top 7 of causes of occupational accidents						
	A	B	C	D	E	F	G
Austria (A)	-	-	-	3	-	4	-
Belgium (B)[1]	2	5	1	-	4	3	-
Denmark (DK)	1	-	-	3	5	-	-
Spain (E)	2	3	1	-	-	5	4
Finland (FIN)	3	(2)[2]	1	-	-	4	5
Germany (GER)	(3)[3]	4	-	-	-	-	-
Ireland (IRL)[7]	-	(4)	-	(1)	(2)	-	(3)
Luxembourg (L)	-	1[4]	3[4]	5[4]	-	-	4[4]
Netherlands (NL)[5]	-	-	-	-	-	-	-
Sweden (S)	1	2	5	4	3	-	-
United Kingdom (UK)	(1)[6]	2	3	-	-	-	-

A = Overload/extension of the body; B = Slips,trips or falls on the same level; C = Struck by object or tool; D = Cut/stabbed by sharp objects;
E = Tripped/pinched/crushed; F = Falling objects; G = Fall from a height

Procedure for the construction of this table:
Eleven countries have provided a list of causes of occupational accidents. For each country a top 5 of main causes has been composed. When these causes are aggregated the following ranking of 7 causes over 11 countries appears.
For instance, category A ('overload/extension of the body') appears to be the highest ranking on average in six countries (B,DK,E,GER,S,UK). Category B ('slips,trips or falls on the same level') appears to be on average the second highest ranking, etc.

Remarks:

1) B Regarding this category data are also available according to economic sector.
2) FIN the category 'accidents due to stumbling and slipping' has the second highest score regarding the number of occupational accidents.
3) GER the category 'manual transport' has been added to the category 'overload/extension of the body'.
4) L the causes of accidents refer to reported occupational accidents within the sector 'agriculture and forestry'. The second cause of accidents in this sector ('accidents caused by animals') was not taken up in this table, because it did not fit into the top 7 of causes among all eleven countries. Data on accidents in the industrial sector are also available. However these data cannot be included into this table due to the fact that different categories were used.
5) NL data on causes of accidents have been provided. However the categories do not correspond with the categories of the matrix (and the other countries), so that they could not be fitted in the table.
6) UK the category 'handling,carrying and lifting' has been added to the category 'overload/extension of the body'.
7) IRL the figures for Ireland were deemed highly unreliable.

Causes of occupational accidents:
An inventory of data provided by eleven countries (A,B,DK,E,FIN,GER,IRL,L,NL,S,UK), made it possible to compile an overview of a top 7 of causes of accidents (see table 5.2.1). Although the Netherlands have provided data on causes of accidents, the categories could not be fitted into this table.
In six countries (B,DK,E,GER,S,UK) **'overload/extension of the body'** (in table 5.2.1b depicted as 'A') is an important cause of accidents. In Denmark, Sweden and the United Kingdom this diagnosis is the most frequent. In Belgium and Spain it is in second place and in Germany it is in third place regarding number of accidents.
Another important cause of accidents is **'slips, trips, falls on the same level'** ('B' in table 5.2.1b). In eight countries (B,E,FIN,GER,IRL,L,S,UK) this category is a major cause of accidents. In Luxembourg it is for the agricultural sector cause of accidents number 1.
In three countries (FIN,S,UK) it is responsible for the second highest number of all accidents. In Spain it is in third place, in Germany and Ireland it is the fourth and in Belgium it is the fifth highest cause of accidents.

The other causes of accidents (from the top 7 from table 5.2.1b) that have been reported:

(C) **being struck by an object or tool:** Belgium(1st), Spain(1st), Luxembourg(3rd) Sweden(5th) and the United Kingdom(3rd);

(D) **being cutted/stabbed by sharp objects:** Austria(3rd place), Denmark(3rd), Ireland(1st) Luxembourg(5th) and Sweden(4th);

(E) **being tripped, pinched or crushed:** Belgium(4th), Denmark(5th), Ireland(2nd) and Sweden(3rd);

(F) **falling objects:** Austria(5th), Belgium(3rd), Spain(5th) and Finland(5th)

(G) **fall from a height:** Spain(4th), Ireland(3rd) and Luxembourg(4th).

Economic sectors:
More in depth study on ranking causes of occupational accidents within sectors is possible. However within the framework of this report this could not be carried out due to time limits.

Conclusion on occupational accidents:
Some remarkable differences in figures between countries have emerged. Further analysis of the methodologies used to compile these figures is however needed to reliably draw conclusions.

Table 5.2.2: ***Data on sickness absenteeism in 10 European countries#***

Parameters / Countries	SICKNESS ABSENTEEISM IN GENERAL				
	Lost days	Total of cases	Volume	Frequency	Average duration of spell
Austria (A)[1]	-	3,039,663	-	-	-
Belgium (B)[2]	-	-	5.8%	0.7	13.6
Denmark (DK)[3]	-[3]	-	-[3]	-	-
France (F)[4]	-	-	5.3%	-	-
Finland (FIN)[5]	16,633,000	-	8.2%	-	-
Germany (GER)[6]	-	-	-	1.6	15.8
Luxembourg (L)[7]	1,192,058	118,282	-	-	10.1
Netherlands (NL)[8]	-	7,614,000	8.1	1.6	-
Sweden (S)[9]	-	219,400	-	-	-
United Kingdom (UK)[10]	627,242,400	-	-	-	-
Range of total	**1,192,058(L) - 627,242,400(UK)**	**118,282(S) - 7,614,000(NL)**	**5.3%(F) - 8.2%(FIN)**	**0.7(B) - 1,6(NL;GER)**	**10.1(L) - 15.8(GER)**

Sickness absenteeism in general: volume = total number of calender days lost due to sickness as % of the total number of calendar days per year; frequency = the average rate of sickness reports; average duration of spell = the average number of days of a period of absence, due to sickness.

See introduction

Remarks:

Sickness absenteeism in general:

1) A: the total number of cases refer to the year 1993.
2) B: there is no systematic registration of sickness absenteeism in Belgium. The figures on sickness absenteeism derive from a survey (2,778 employees in 11 institutions) that was done by the Employers Interfactory Occupational Health Service (IDEWE) in 1994. The rate (frequency) of sickness reports is low, because there is a great probability that, within that period, cases have been forgotten.
3) DK: data was provided, but deemed highly unreliable, because of problems with estimation of the fraction of part-time employment.
   (1) the information does not refer to persons with so called "loose affiliation" (persons who have been employed very shortly or who are chronically ill) and to self employed persons;
   (2) the percentages are calculated on the basis of calender days;
   (3) the absence period less than 3 days is not calculated within the total number of lost calender days.
4) F: in France there is no systematic registration of sickness absenteeism. Five major surveys have been done since the Second World War. The last was in 1991 and included 62,000 companies.
5) FIN: N = the number of lost sickness days; the percentage refers to the number of lost sickness days divided by the number of employed workers of wage earners in the concerned category.
6) GER: the statistics comprise all data relating to the compulsory insured in Germany. Very detailed

statistics are taken up in the report. However, it is difficult to extrapolate the basic figures for absenteeism (total percentage, frequency and average length).

7) L: the figures refer to the year 1992.

8) NL: the volume of sickness absence is calculated on the basis of working days (as opposed to calender days). The total number of absent insured working days is expressed as a percentage of the total number of insured working days (in a year).

9) S: the data were extracted from the Swedish Labour Force Surveys (1993). The number of registered cases refer only to the employed.

10) UK the data on sickness absenteeism came from the Department of Social Security and is based on a sample 1% of claims and/or Invalidity Benefit in Great Britain for the period 06.04.92 - 03.04.93.

Sickness absence in general

Table 5.2.2 shows that ten countries (A,B,DK,F,FIN,GER,L,NL,S,UK) have provided data regarding sickness absenteeism in general. The data of these countries are dispersed over the five parameters. It is not possible to give a complete picture of figures from the various countries. The data can be summarized as followes:

Lost days: three countries provided data (FIN,L,UK). The range of the total of lost days varies from 1,192,058(L) to 627,242,400(UK).

Total of cases: four countries provided data: Luxembourg(L): 118,282 cases, Sweden(S): 219,000 cases, Austria(A): 3,039,663 cases and the Netherlands(NL): 7,164,000 cases.

Volume: four countries (B,F,FIN,NL) provided data. The range of percentage of lost days varies from 5.3%(F) to 8.2%(FIN).

Frequency: three countries (B,GER,NL) provided data. The range of rate of sickness reports varies from 0.7(B) to 1.6(GER,NL).

Average duration of spell: three countries provided data. In Luxembourg the average duration is 10.1 days; In Belgium it is 13,6 days and in Germany it is 15.8 days.

Conclusion: there is a lack of comprehensive data in the various European countries regarding sickness absenteeim in general, so it is difficult to compare figures between various countries .

Gender:

Nine countries (A,B,DK,F,FIN,GER,NL,S,UK) have provided data according to gender. However there is not enough data on one of the parameters, mentioned in the table above, therefore the figures will not be described in this report.

Age:

Seven countries (B,FIN,GER,L,NL,S,UK) provided data on different parameters of sickness absenteeism according to age-categories. Not enough data on any of the parameters was available to make a description; therefore this issue will not be handled in this report.

Table 5.2.3: *Number and rate of reported occupational diseases in 1 year (all cases and diagnosis) in 12 European countries*

Countries*** / Item		REPORTED OCCUPATIONAL DISEASES# All cases*	Top 6 of diagnosis** A	B	C	D	E	F
Austria (A)	N / rate	2,100[1] / 58	2	-	1	3	4	-
Belgium (B)	N / rate	2,367[2] / 922	3	-	-	5	1	-
Denmark (DK)	N / rate	15,655[3] / 600	3	1	2	5	-	-
Spain (E)	N / rate	5,489[4] / 46	2	4	3	-	-	-
Finland (FIN)	N / rate	7,006[5] / 319	4	1	3	5	-	2
Germany (GER)	N / rate	92.058[6] / 249	2	-	-	3	1	-
Italy (I)	N / rate	18,037[7] / 85	3	-	1	4	-	2
Ireland (IRL)	N / rate	(83*)[8] / (7*)	(2*)	(1*)	-	(3*)	-	-
Luxembourg (L)	N / rate	146 / 74	1	2	4	3	-	-
the Netherlands (NL)	N / rate	1,086[10] / 18	-	-	-	-	-	-
Sweden (S)	N / rate	33,773[11] / 852	4	1	3	5	-	-
United Kingdom (UK)	N / rate	[12]	-	-	-	-	-	-
Range of total	N	146(L) - 3,300(S)	A = Skin diseases; B = Musculo-skeletal disorders; C = Hearing disorders; D = Respiratory diseases; E = Physical agents; F = Asbestosis-induced diseases					
	rate	18(NL) - 852(S)						

\# See introduction

*) All cases: N = number of occupational diseases; rate = (number of occupational diseases : the total working population) x 100,000.

**) Diagnosis (A-F): 1 = most frequent reported occupational disease in the country, to 5 = fifth most frequent reported occupational disease in the country.

***) Data on France was received too late to be taken up in the table.

Eleven countries have provided a list of causes of occupational diseases. For each country a top 5 of main causes has been composed. When these causes are aggregated the following ranking of 6 causes over 11 countries appears.

For instance, category A ('skin diseases') appears to be the highest ranking on average in ten countries. Category B ('musculo-skeletal disorders') appears to be on average the second highest ranking, etc.

Remarks:

1) A: it is not clear whether the number of diseases refers to reported or to notified diseases. The total number of occupational diseases refer to the year 1993. The total working population: 3,608,100.
2) B: the total number of occupational diseases refers to the year 1993. The total working population: 2,573,749 (excl. 600,000 employees from governmental and educational sector).
3) DK: the rate is based on a total working population of 2,609,859 (1992).
4) E: the total number of occupational diseases refers to the year 1993. The total working population of that year was 11,838,000.
5) FIN: the rate is based on a total working population of 2,199,000 (1992).
6) GER the total number of occupational diseases refers to the year 1993. The total working population of 36,940,000 derives from 1992 (ILO, 1994b).
7) I: the rate is based on a total working population of 21,145,00 (1989).
8) IRL: data regarding reported occupational diseases are very scarce. There is a significant under-reporting on this subject. Nevertheless, the rate is based on a total working population of 1,139,300 (1993).
9) L: the total number of reported occupational diseases refer to the year 1993 and is based on a national working population of 196,800 (1992).
10) NL: the total number of reported occupational diseases refers to 1993. In the Netherlands the available data on occupational diseases are not very reliable due to significant underreporting.
11) S: persons in the working population covers all persons who are either gainfully employed or unemployed. The rate is based on a total working population of 3,964,000 (ILO, 1994b).
12) UK: figures on perceived work related ill-health as reported by the work force are available. However the fraction of occupational diseases in these figures cannot be distinguished.

General remarks:

The figures are based on a total working population (i.e. the employed and self-employed), except Belgium (600,000 employees are excluded).

Table 5.2.3 shows that eleven countries (A,B,DK,E,FIN,GER,I,IRL,L,NL,S) have provided data regarding the category 'reported occupational diseases' and the same countries (except Austria) have provided data on the category 'diagnoses of reported occupational diseases'.

Reported occupational diseases:

The range of total number of reported occupational diseases varies from 146(L) to 92,058 (GER). Relatively high rates are reported in Denmark, Sweden and Belgium respectively 600, 852 and 922 per 100,000 workers. Low rates as in e.g. Ireland and the Netherlands are indicated to suffer from substantial under reporting.

Gender:

Six countries (B,DK,E,FIN,I,S) have provided data according to gender. In two countries (DK,S) the ratio male/female workers on occupational diseases are the same. In four countries (B,E, FIN,I) the percentage of male workers affected by an occupational disease is somewhat higher than the percentage of female workers. However these differences are very small. Finally, Germany has indicated that there are data on gender available. However the figures are not taken up into the German report.

Age:
Six countries (B,DK,E,FIN,GER,I) have provided data according to age. The age categories among these countries differ however too much (only Belgium and Finland, and to some extent Denmark have used the same categories in their reports), therefore no comparisons on this issue are made.

Diagnoses of reported occupational diseases:
An inventory of data provided by ten countries (A,B,DK,E,FIN,GER,I,IRL,L,S) made it possible to compose an overview of a top 6 of diagnoses of reported occupational diseases (see table 5.2.3). In these ten countries **'skin diseases'** (depicted as 'A' in table 5.2.3) is the most important diagnosis of reported occupational disease. In Luxembourg it is diagnosis number one. In Austria, Spain, Germany and Ireland it counts for the second highest diagnosis category regarding the number reported occupational diseases. Finally, in Belgium, Denmark and Italy it is on third place and in Finland and Sweden on fourth.
In six countries (DK,E,FIN,I,L,S) **'musculo-skeletal disorders'** (category 'B') is another major category of reported occupational diseases. In Denmark, Finland, Ireland and Sweden, this diagnosis has the highest number of reported occupational diseases; Luxembourg (2th) and Spain (4th).
Seven countries (A,DK,E,FIN,I,L,S) provided data indicating that **'hearing disorders'** (category 'C') is an important occupational disease. In Austria and Italy it is the most common reported occupational disease; in Denmark (2th), in Spain, Finland, Luxembourg and Sweden (3th) and in the United Kingdom (4th).

<u>The other diagnoses of reported occupational accidents (see top 6 from table 5.2.3) that have been reported:</u>

(D) respiratory diseases: Austria(3rd), Belgium(5th), Denmark(5th), Finland(5th), Germany(3rd), Italy(4th), Ireland(3rd), Luxembourg(4th), Sweden(5th) and the United Kingdom(2nd);
(E) physiological agents: Belgium(1st), Germany(1st), Austria (4th);
(F) asbestosis-induced diseases: Finland(2nd), Italy(2nd) and the United Kingdom(5th). The high figure from Finland is due to the large number of pleural fibrosis and calcification identified in a comprehensive screening project.

Conclusion on occupational diseases:
Since the definitions of occupational diseases, the reporting, and the compensation systems vary considerably among the various countries, the cross-national comparability of numbers and rates is poor. Further analysis is indicated to reliably draw conclusions.

Table 5.2.4: ***Number and rate of cases of occupational mortality in 8 European countries.#***

Countries*		Occupational mortality (all cases)
Belgium (B)	N rate	1,009 27[1]
Denmark (DK)	N rate	39,671 1486[2]
Finland (FIN)	N rate	139 6[3]
France (F)	N rate	1,415 10[4]
Italy (I)	N rate	1,937 9[5]
the Netherlands (NL)	N rate	59 1[6]
Sweden (S)	N rate	159 4[7]
the United Kingdom (UK)	N rate	430 2[8]
Range of total	N rate	**59(NL) - 39,671(DK)** **1(NL) - 1,486(DK)**

\# See introduction
N = the number of death cases; rate = (the total number of cases / the total working population) x 100,000.
*) Data from Austria was received too late to be taken up in the table.

Remarks:

1) B: the total number of occupational mortality refers to the year 1993. The total working population: 3,745,500 (ILO,1994). The total number consist of the total number of death cases due to fatal occupational accidents (271) and occupational diseases (738).
2) DK: the rate is based on a total working population of 2,670,000 (1990).
3) FIN: the rate is based on a total working population of 2,199,000 (1992). The total number refers to the total number of death cases due to fatal occupational accidents (64) and asbestos related diseases (75).
4) F: the rate is based on a total working population of 14,440,400 (1992). The total number refers to the total number of death cases due to fatal occupational accidents (1,172) and occupational diseases (243).
5) I: the rate is based on a total working population of 21,595,00 (1991). The total number refers to the total number of death cases due to fatal occupational accidents.
6) NL: the rate is based on a total working population of 6,155,000 (1989). The total number refers to the total number of death cases due to fatal occupational accidents.
7) S: the rate is based on a total working population of 3,964,000 (1993). The total number refers to the total number of death cases due to exposure to safety risks.
8) UK: the rate is based on a total work force of 25,728,000 (1992). The total number refers partly to the total number of death cases due to exposure to safety risks.

Occupational mortality

Table 5.2.4 shows that eight countries provided data regarding all cases of occupational mortality. The picture that emerges is quite diverse. In four countries (B,DK,FIN,F) the data refer to occupational accidents and diseases and in two other countries (I,NL) only to occupational accidents. Finally in two countries (S,UK) the data refer to exposure to safety risks. When looking at the figures as such, those from Denmark are remarkably high and claimed to be reliable. This result cannot be explained from major differences in methodology between countries.
Data on occupational mortality is also available in Austria, but this was received too late to be taken op in the table.
In the report of Germany and Luxembourg there are no figures given on occupational mortality. However a reference has been made to the figures regarding the number of fatal accidents and occupational diseases.

Table 5.2.5: ***Number and rate of people regarding general mortality in 11 (out of 16) European countries***

Countries		GENERAL MORTALITY							
		All cases	Top 7 of diagnoses						
			A	B	C	D	E	F	G
Austria (A)	N rate	83,162[1] 1068[1]	1	2	3	5	-	4	-
Belgium (B)	N rate	107,336[2] 1081[2]	1	2	5	-	4	-	3
Denmark (DK)	N rate	60,821[3] 1178[3]	-	-	-	-	-	-	-
Spain (E)	N rate	324,796[4] 836[4]	1	2	4	3	-	5	-
Finland (FIN)	N rate	49,852[5] 991[5]	1	2	3	4	-	5	-
Greece (GR)	N rate	95,498[6] 944[6]	1	2	5	4	3	-	-
Italy (I)	N rate	544,500[7] 943[7]	1	2	4	3	-	5	-
Luxembourg (L)	N rate	3,916 1,035[8]							
Netherlands (NL)	N rate	129,887[9] 859[9]	1	2	4	3	5	-	-
Sweden (S)	N rate	94,226[10] 1,091[10]	-	-	-	-	-	-	-
United Kingdom (UK)	N rate	570,044[11] 987[11]	-	-	-	-	-	-	-
Range of total	N rate	**3,916(L) - 570,044(UK)** **836(E) - 1,178(DK)**	A = diseases of the circulatory system; B = neoplasms; C = injury and poisoning; D = diseases of the respiratory system; E = symptoms, signs and ill-defined conditions; F = diseases of the digestive system; G = other diseases						

N = number of reported death cases; rate = (number of registered death cases in the reference period / the national population) x 100,000.

Seven countries have provided a list of diagnoses related to general mortality. For each country a top 5 of main diagnoses has been composed. When these diagnoses are aggregated the following ranking of 7 diagnoses over 7 countries appears.
For instance, category A ('diseases of the ciculatory system') appears to be the highest ranking on average in seven countries. Category B ('neoplasms') appears to be on average the second highest ranking, etc.

Remarks:

1) <u>A:</u> the rate is based on a national population of 7,860,800 (1992).
2) <u>B:</u> The number of cases refer to the reference year 1989. For fixing the rate a national population of 9,927,600 has been used.
3) <u>DK:</u> to fix the rate an estimated national population of 5,162,100 (1992) has been used.
4) <u>E:</u> the number of cases refer to the year 1989. For fixing the rate a national population of 38,851,900 has been used.
5) <u>FIN:</u> the number of cases refer to the year 1992. For fixing the rate a national population of 5,029,000 has been used.
6) <u>GR:</u> the rate is based on a national population of 10,120,000 (1991).
7) <u>I:</u> the rate is based on a national population of 57,746,200 (1991).
8) <u>L:</u> the rate is based on a national population of 378,400 (1990). The figures on general mortality refer to the reference period 1986-1990.
9) <u>NL:</u> the rate is based on a national population of 15,129,000 (1992).
10) <u>S:</u> the rate is based on a national population of 8,644,100 (1992).
11) <u>UK:</u> the rate is based on a national population of 57,749,000 (1992).

General mortality
Table 5.2.5 shows that eleven countries (A,B,DK,E,FIN,GR,I,L,NL,S,UK) have provided data regarding the categories 'all cases' and 'diagnoses of general mortality'.

All cases:
The range of total number of death cases varies from 3,916(L) to 570,044(UK). The range of rates of the ten countries varies from 636(E) to 1178(DK) death cases per 100,000 inhabitants. Relatively high rates are reported in Austria, Belgium, Denmark, Luxembourg and Sweden, respectively 1068, 1081, 1178, 1035 and 1091 death cases per 100,000 inhabitants. For the remaining countries the rates are below 1000. Relatively low rates are reported from Spain and the Netherlands, respectively 836 and 859 death cases per 100,000 inhabitants.

Gender:
Ten countries (A,DK,E,FIN,GER,GR,I,L,NL,S) have provided data according to gender. In all countries (except Austria and Denmark) the percentage of male inhabitants that has died in 1 year is somewhat higher than the percentage of female inhabitants. The differences between male and female inhabitants are however very small. Only Denmark reports that the percentage of death cases for male and female inhabitants are the same.

Age:
Nine countries (A,B,DK,FIN,GER,GR,I,L,S) indicated that data according to age categories were available. However eight countries (A,B,DK,FIN,GR,I,L,S) actually provided information in their national reports. Due to the fact that the age categories among these eight countries differ too much, it is not possible to make any meaningfull remarks about the data.

Diagnoses of general mortality:
An inventory of data provided by seven countries (A,B,E,FIN,GR,I,NL), made it possible to compose an overview of a top 7 of diagnoses concerning causes of death (see table 5.2.5). In all seven countries the diagnosis **'diseases of the circulatory system'** is the first death cause .The category **'neoplasms'** is cause of death number 2. **'Injury and poisoning'** is also an important cause of death within all seven countries. In Austria and Italy it is cause of death number 3, in Spain, Italy and the Netherlands it is cause of death number 4 and in Belgium and Greece it is

cause of death number 5.

Six countries (A,E,FIN,GR,I,NL) reported that **'diseases of the respiratory system'** is another major cause of death. In Spain, Italy and the Netherlands it is the thirds highest cause of death; in Finland and Greece it is on number 4 and in Austria number 5.

The other causes of death (from table 5.2.5) that have been reported:

(E) **symptoms, signs and ill-defined conditions:**	Belgium(4th place), Greece(3rd) and the Netherlands(5th);
(F) **diseases of the digestive system:**	Austria(4th), Spain(5th), Finland(5th) and Italy(5th);
(G) **other diseases:**	Belgium(3).

Gender:
Six countries (A,E,FIN,GER,I,NL) indicated to have data according to gender. Due to the fact that German report did not contain actual data on this subject, this issue will not be included in this report.

Age:
Seven countries (A,B,E,FIN,GER,I,NL) indicated to have data according to age. However, six countries (A,B,E,FIN,I,NL) actually provided information in their national reports. Due to the fact that the age categories among these six countries differ too much, it is not possible to make any meaningful remarks about the data.

5.3 Discussion and preliminary conclusions

The discussion and conclusions hereafter are based on the previously presented data provided in the national reports from 16 countries, and on information from the first European Survey on the Working Environment (European Foundation for the Improvement of Living and Working Conditions, 1992).

It should be noted however, that a more thorough weighing of the presented data with regard to their reliability and comparability, as to come to justified conclusions, will take place in chapter 7 where the overall discussion and final conclusions will be presented. In general can already be said though, that the conclusions hereafter must be considered as indicative. This is due to cross-national differences in methodology, but also because the data on the working environment are mostly 'subjective' (i.e. based on questionnaire-based surveys), whereas the data on health and safety output often are known or assumed to be subject to significant underreporting (see section 4.3). Furthermore, it is noted that comparison of the data from the national information sources, with data from the Eurosurvey can only be done at a global level. This may pinpoint some remarkable findings, which could be interesting for future in-depth analysis.

5.3.1 Working environment

Conclusions on the *working environment across Europe*:
To identify main risk factors in the working environment across Europe, i.e. those risk factors with the largest population exposed, three ways can be taken. Firstly by ranking the average percentages of exposed, as presented in the last line of tables 5.1.1 - 5.1.3. Secondly by

composing a 'top 3' of risk factors in each country and look for cross-national overlap. The third way is to evaluate the outcomes of the first European Survey on the Working Environment. The first two ways of analysis are impeded by cross-national differences in the methodology used to assess the number of people exposed, as well as by the fact that six countries (GR, IRL, I, L, P, UK) have no national information sources on the working environment at all, whereas two countries (B, F) provided data from national information sources only on physical risk factors, and not on the psychosocial and physiological risk factors. The third way is hindered by the fact that it covers less risk factors and partly other countries, than the other two ways. Yet, an overlap in the 'top 5' from all three ways could provide the best indication for main risk factors in the European working environment, and is therefore presented here.

The first way of analysis leads to the following conclusions:

- Aggregation of the data from national sources into weighted European averages of exposed people indicate three main risk factors in the working environment across Europe:
  * working pace: an average of at least 51% of the European working population is exposed;
  * job content: at least 45% is exposed;
  * lack of influence and control over own work: at least 39% is exposed.
- At least 24-32% of the European working population is exposed to hazards concerning:
  * working postures and movements ($\geq$ 32%);
  * working hours ($\geq$ 29%);
  * climate ($\geq$ 26%);
  * manual handling ($\geq$ 25%);
  * noise ($\geq$ 24%).
- Least common risk factors regard vibrations (7%) and radiation (2%), whereas the position of social interaction in this ranking of risk factors could not be determined, because no European average of exposed workers could be calculated.
- Hence, according to this way of analysing, three risk factors in the psychosocial work environment pose the most problematic areas across Europe, while factors of the physical working environment and physiological factors do so to a lesser extent.

The second way of analysis results in the following conclusions:

- The overall most frequently appearing risk factors in the 'top 3' of the various countries that provided data from national sources, are:
  * job content;
  * working pace;
  * social interactions;
  * influence and control over own work.
- Less frequent risk factors in national 'top 3' are strenuous working postures and movements, noise, vibrations and radiation. Working hours and climate appear in only 1 country in the 'top 3', and manual handling in none of the countries.
- Hence, also according to this second way of analysing, risk factors in the psychosocial work environment can be considered as main risk factors, whereas some physical and physiological loads are also frequently present in national 'top 3'.

Conclusions from the third way of analysing are as follows:

- Ranking the risk factors according to the average percentages of exposed workers in the twelve countries that were involved in the Eurosurvey, indicate the following 'top 5' of risk factors:
  * job content (51%);
  * working pace (49%);

* strenuous working postures and movements (42%);
* influence and control over own work (38%);
* climate (34%).

- Lesser widespread risk factors regard manual handling (31%), noise (27%), social interaction (18%) and working hours (18%).
- Comparisons of these average percentages of exposed workers in the various countries, with those in the first way of analysing, can only be indicative, because different sub-items per risk factor in each country may be involved. Yet, the average European percentages from the Eurosurvey and those based on the national data are in any case interesting to put next to each other, because they both indicate the proportion of workers who report to be exposed to a certain aspect of a risk factor. The result of this comparison is a rather miscellaneous picture. The percentages of workers exposed to working pace, lack of influence and control over own work, and to noise are more or less equal. For job content, manaul handling, climate and strenuous working postures lower percentages are found in the Eurosurvey, than by the aggregation from the national sources (6-10% lower). Only for working hours the percentage from the Eurosurvey is higher (10%). On vibrations, radiation and social interaction no such comparisons could be made, because only from one of the two sources average percentages are available.

The *overall conclusions on the working environment across Europe*, based on the three ways of analysing the available data, consequently are:

- Main, i.e. most widespread, risk factors are:
  * job content;
  * working pace;
  * lack of influence and control over own work;
  * strenuous working postures and movements.

  These hazards appear in the 'top 5' resulting from all three ways of analysing.
- No consistency exists with respect to noise, climate, working hours and manual handling which all take various positions in the rankings resulting from the three ways of analysing. This also counts for social interaction, vibrations and radiation, for which only two ways of analysis are available, however.

Conclusions on the *working environment in the various countries*:

- Within the framework of the project, the ranking of countries according to highest or lowest percentages of exposed workers was not feasable, because of the cross-national differences in methodology by which these data were obtained, which largely influence their comparability. Yet, secondary analysis of the available data remains a distinct possibility for future research.
- The same conclusion counts for the comparison of the data from national sources with the information from the Eurosurvey.

Conclusions on *main risk groups across Europe*:

- In general it is concluded that main risk groups with respect to the work envrionment can be identified only to a very limited extent. Firstly, because data according to age and occupation were not sufficiently available for a proper analysis (for all risk factors from only less than six countries). Hence, risk groups cannot be identified among age groups and occupational groups.
- Risk groups with respect to gender can however be identified, but only with respect to two risk factors: noise and vibration. On the other nine risk factors not enough countries provided data on comparable sub-items to justify analysis.
- In all countries that provided data on exposure to noise and vibrations according to

gender, the percentage of exposed male workers is circa 10-20% higher than the percentage of female workers.

Conclusions on *most hazardous sectors across Europe*:
- Hazardous sectors in Europe cannot be identified from the obtained data either, because not enough countries (less than 6) provided comparable data.

5.3.2 Health and safety output

On 5 output-items or indicators sufficient data were available to report. These indicators are:
- occupational accidents;
- sickness absence in general;
- occupational diseases;
- occupational mortality;
- general mortality.

In the previous section the tables on these indicators were presented and some interesting features were highlighted. The question now arises as to what inferences are to be drawn from this information. First of all it may be useful to view the types of indicators available on a larger scale for this analysis. From the 5 indicators only 3, i.e. occupational accidents, diseases and mortality are actually directly indicative for the health and safety output within the framework of the working environment. The remaining 2 indicators, i.e. sickness absence and general mortality are merely indirectly related to the working environment.
Furthermore can be observed that with regard to occupational diseases and occupational mortality the output in the majority of countries is extremely limited in terms of actual cases. Only with regard to occupational accidents substantial numbers can be found, both in the category 'all accidents' and 'accidents with work interruption'.

Occupational accidents: all accidents, accidents with work interruption:
Within these respective categories both Spain (903, 508) and Belgium (897, 611) show especially high rates per 10,000 workers. Luxembourg reported a particular high rate (1,002) for 'all accidents', and France and Portugal for 'accidents with work interruption' (519, respectively 546). These outcomes cannot be attributed to differences in definitions used (e.g. one or three days of absence to register as an occupational accident. In all countries that provided information on the issue *gender*, the males showed higher rates of 'accidents with work interruption'. As far as *economic sector* is concerned, it appeared that especially (metal)manufacturing industries and building/engineering sectors showed relatively high rates.

Occupational accidents: fatal accidents:
Also for the category 'fatal accidents' Luxembourg (1.2), Belgium (1.1) and Spain (1.0) show the highest rates.
Within the category 'fatal accidents' an overall higher rate could be observed for male workers, especially in Denmark, Spain, Germany and Italy. Italy reported relatively high rates in the agriculture and building sector and Spain in the sectors energy and water, building and transport.

Occupational diseases:
It is clear from the data on occupational diseases that there are major differences between countries as far as the levels of reported cases is concerned. It appears that in general the conclusion can be drawn that in countries with a more sophisticated infrastructure with regard to data gathering (and defining of diseases) the level of reported cases is significantly higher than in

countries lacking this infrastructure. This means that a shallow comparison of the presented data would lead to false conclusions. In the case of occupational diseases also significant under reporting is assumed in several countries.
As far as the diagnoses of occupational diseases are concerned, especially 'skin diseases', musculo-skeletal disorders' and 'hearing disorders' appear to be predominante.

Occupational mortality:
As far as occupational mortality is concerned the conclusion is justified that the data is probably of insufficient quality. Except for Denmark where a claim is put on reliability, the information from the other countries appears highly unreliable and at least of diverse nature.

Sickness absenteeism:
In the case of sickness absence data it is remarkable that in virtually no country, with the exception of Finland, Sweden and the UK, data was available on occupational sickness absence. The information gathered on general sickness absence during this research is rather diverse in nature and is therefore difficult to compare. Indicative in this context is the fact that only 4 countries were able to produce sickness absence percentages. Although other research (European Foundation for the Improvement of Living and Working Conditions, 1994d) has produced more data on this specific subject.

General mortality:
The data on general mortality seems to be fairly consistent. The data as such does not seem to suffer from under reporting, therefore ensuring a high level of reliability and comparability. Here Denmark is at the top of the list with a rate of 1178 (per 100.000) followed by Sweden (1091), Belgium (1081) and Austria (1068). Relatively low rates are found in Spain (836) and the Netherlands (859). Furthermore, in all seven countries that provided data on mortality according to diagnoses there is consistency over the two main death causes, being diseases of the circulatory system and neoplasms.

The overall conclusion is therefore that in the actual monitoring of the working environment in terms of health and safety output only occupational accident data is sufficiently available. The fact that in most countries data on occupational accidents is available probably stem from the fact that in most countries occupational accidents are related to compensation schemes. Furthermore, accidents at work are easier to determine and to define for international comparison, than for example occupational diseases. Although it must be noted that in several countries under reporting seems to be a significant problem, both for occupational accidents and occupational diseases.

A further question could be whether there is some comparability between the countries with regard to the causes of the output indicators. Although in some countries the list of causes has sometimes been slightly altered in general, a reasonable effective comparison can be made between countries. However from the 15 countries that provided data on occupational accidents 4 countries, i.e. Greece, Italy, the Netherlands and Portugal, were not able to produce data on causes or - in the case of the Netherlands - used a deviating system in classifying causes.
Generally speaking no significant differences can be found between countries with high rates and countries with relatively low rates with regard to the causes of occupational accidents. Especially the categories 'overload/extension of the body' and 'slips, trips and falls' seem to be predominant causes throughout Europe.

6 *Trends and strategies in Europe regarding data production*

Introduction

In this chapter the aim is to present an qualitative overview of trends and strategies regarding data production in Europe. For descriptive purposes a distinction has been made between 3 topics:
1. Issues that have been stated as main problem(s);
2. Topics that have been chosen to take initiatives upon;
3. Developments and/or activities that are taking place regarding data gathering and production and types of approach in these developments (e.g. sectorial).

For each of these issues data has been gathered, with a special emphasis on the type of 'actor' involved (government, social partners or other organizations). In the following paragraphs the topics have been described (mostly in general) based on the information received. In general the principal 'actor' involved is the government. Wherever relevant remarks could be made about other 'actors' were they included in the paragraphs.
No data were obtained from Italy, Greece, Portugal, France and Norway, and due to time limits, information on Luxembourg has not been included.

6.1 Identified problems and main topics in governments', social partners', and other authoritative organizations' policies

In screening the available information of the participating countries it became clear that a wide variety of problems and/or topics have been identified. The information received regarding issues that are identified as the main problems involves 8 countries, i.e. Belgium, Denmark, Ireland, Finland, Germany, The Netherlands, Spain and the United Kingdom. These same countries also provided information regarding topics on which initiatives have been taken. Firstly 'the main problems' are dealt with by looking on the one hand for common ground and on the other hand for the most interesting exceptions to the rule. After that 'the initiatives' are dealt with accordingly.

Identified main problems
In analysing the available data 2 main categories of problems can be distinguished. Firstly there are several countries in which the lack of information is mentioned as the predominant problem. Specifically this concerns information in general on causes and effects of working conditions. The fact that there is no centralized institution commissioned to deal with information gathering is mentioned as an important cause for this lack of information. Although in some countries some data may be available, the reliability and comparability is considered poor due to the lack of a clear structure in data gathering. Difficulties in gaining access to this data are also mentioned as a problem.
In practically all of the 10 countries the government is the main actor as far as stating the lack of information as the 'main problem' is concerned. The social partners do not seem to play an overactive role in this. Although in Belgium particular concern was expressed about the inadequacy of statistics on accidents by the social partners. Furthermore here the difficulties concerning information gathering on a European level were underlined. Also authoritative organizations in Belgium mentioned the problems concerning inadequate statistical information.
Unspecified sources in Belgium, Germany and the United Kingdom refer especially to difficulties in obtaining information from the smaller and medium sized large companies. Again this refers to a lack of adequate organization in the data gathering process.

In the second place a wide variety of health and safety issues are considered to be especially problematic. The Spanish government in 1992 mentioned e.g. accidents, noise, use of dangerous substances, heavy loads, increasing new hazards such as the use of computers, lower autonomy and repetitive tasks. The Dutch labour directorate as well as the largest Dutch Labour Union (FNV) stated e.g. psychological and physical exposure as main problems. Furthermore due to changes in legislation with regard to sickness benefit insurance arrangements a major problem in data gathering in this field has occurred in the Netherlands. Of particular concern to the Irish government is safety management, in relation to the implementation of the European Directives; and in particular manual handling, personal protective equipment, first aid and accident reporting. In the United Kingdom concern was expressed by all parties with regard to maintaining high standards of health and safety through inspection, guidance, advice, enforcement and research. Furthermore the need to ensure that employers receive clear messages in applying risk assessment and the cost effectiveness of good health and safety performance were mentioned as main problems. Also privatisation and technological change are regarded as problem areas. In Finland the main problems were stated as: severe new types of allergies, musculo-skeletal disorders, long-term work disability and premature mortality risk.
In Belgium the limited scope of data on occupational accidents and diseases is considered a problem. Certain categories of the work force are not included in the statistics due to insurance arrangements. Also in Belgium the data on exposure in companies is limited due to the fact that these data are solely based on medical reports mandatory in the legal framework and not on actual risk evaluation.

Main topics in policy
Defining topics to take initiatives on appears mainly a concern of the governments in the various countries, and they propose a wide range of topics. In some countries these topics are proposed as action-programs with mostly general policy oriented objectives, e.g. in Denmark, Finland and the United Kingdom, with an emphasis on legislation and general topics, although some specific topics are mentioned. In other countries an emphasis on more specific problems can be found, e.g. in the Netherlands, Spain, Ireland and Belgium.

In Denmark a program is currently in action on a 'clean working environment by the year 2005' with the purpose of ensuring effective prevention of the principal working environment problems in all sectors. Workplaces will have to provide a safe, healthy and stimulating framework for creativity, quality and productivity at a competitive basis. Seven objectives are set up across all sectors:
- No fatal accidents caused by working environment factors;
- No occupational exposure to carcinogenic chemicals;
- No occupational brain damage due to exposure to organic solvents or heavy metals;
- No young people must suffer serious injury at work;
- No risk of injury due to heavy lifting and no risk of occupational diseases due to monotonous repetitive work;
- No risk of psycho-social disorders due to the way work is organised;
- No risk of hearing damage due to noisy work.

The program sets the stage for cooperation between the social partners and the other players in health and safety at work in the formulation of sectorial objectives for prevention of other serious problems concerning the working environment.

In Finland a number of national priorities are proposed as topics in the future development of Finnish occupational safety and health policies:
- Ensuring effective prevention and control of traditional occupational hazards.
- Developing quality-oriented work organizations by considering the psychosocial aspects of

work, participation and self-regulation
- Improving the integration of occupational safety and health activities with other activities of the enterprises.
- Developing occupational safety and health in small and medium-sized enterprises, and among the self-employed.
- Strengthening of preventive orientation, working capacity activities and methodologies of occupational health services.
- Founding the occupational health and safety policies and practices on sound scientific basis, and developing research to meet the needs.
- Creating an effective surveillance system, to detect and prevent new occupational hazards and deleterious health outcomes in population of working age.
- Providing advice and support to the ageing workers.
- Developing equity at work in view of gender, working conditions, occupational groups, vulnerable groups, migrant workers and workers belonging to different age groups.
- Developing collaboration between authorities, social partners and professional and research institutions and communities.
- Strengthening international collaboration.

In 1986 the FIOH (Finnish Institute of Occupational Health) started with the development of a method for improving the involvement of workers with their work environment (called TUT-TAVA-Action for better workplaces). It is a practical tool to promote the development of total quality management (TQM) in a company.

In the United Kingdom more emphasis is found on legislation:
- Implementing European legislation.
- Updating and reforming the existing legislation in light of technological change and even more rapidly changing international requirements.
- Consolidation of risk assessment provisions in existing legislation.

Furthermore some specific areas of interest are proposed:
- Updating guidance on the legal framework and publishing it.
- Improvement of public health through the 'Health of the Nation' initiative.
- Ensuring that guidance is available particularly for small and medium enterprises.
- Drawing on the results of research to help safety representatives work more effectively with employers.

The second group of countries, i.e Belgium, Ireland, the Netherlands and Spain has formulated a number of specific topics in which to take initiatives. Spain and Belgium are concerned with the improvement of the inventory and analysis of data on working conditions, and Belgium is especially concerned about the prevention of occupational risks at the company level. Furthermore in Belgium the need to document health and safety hazards of workplaces is underlined. In Spain, special attention is given to the identification of risky workplaces and the establishment of preventive policies and priorities.

In Ireland particular emphasis is placed on assisting small and medium sized enterprises. Also the design of a "Safe to Work" booklet that includes hazard identification and risk assessment as well as a model Safety Statement designed especially for small low-risk employment is planned. The Health and Safety Authority will launch a Health and Safety certificate and will be lecturing on a wide range of training courses. Furthermore the Health and Safety Authority has started investigations in the chemical industry.

In the Netherlands, the government has chosen in 1991 a number of topics in which to take initiatives:
- Psychosocial/physiological exposure: (a) the development and execution of research- and information programs; (b) the execution of inspection projects; (c) the development of

standards, instruments and design rules concerning the work situation; (d) the development of policies for Labour inspectors; (e) the working out of proposals and (f) the development of courses.

- Working hours: the development of: (a) a decree on working hours; (b) a reinforcement- and execution policy concerning safety, health and well-being inspection; (c) a national project concerning the prohibition of children's labour.
- Chemical exposure: (a) the development of policies concerning different kind of chemical agents; (b) the development of threshold values, criteria and measures and (c) the execution of the Law 'Environment Dangerous Substances'.
- Physical exposure: (a) contribution to the development and implementation of EC-directives concerning the risks of physical exposure; (b) development of policies on sound, vibrations, radiation, climatological environment and microbiological factors; (c) the development, standardization of measures and conditions for licences.
- Major risks: 'Occupational Safety Reports' are developed as a legal instrument to prevent, minimize and control risks at work.
- Dangerous work: (a) efforts are made for a safer working environment; (b) employers are obliged to impart their employees with knowledge and skills concerning safe working.

Apart from the government also the Dutch Labour Union (FNV) formulated in 1992, several topics to take initiatives on:

- Psychological/physiological exposure: (a) definitions of agreements concerning 'physiological exposure' are inserted in several collective labour agreements; (b) the development of a policy concerning autonomy and participation.
- Working hours: a) the development of a policy concerning the extension of flexible working hours, appropriate working time and rest-breaks.
- Chemical substances/biological materials and physical exposure: a) the stimulation of the development of norms and directives concerning exposure; (b) to plead for the obligation of these norms and directives by the government.

6.2 Developments and activities in countries regarding data gathering and production

6.2.1 General developments

In some countries comprehensive systems to monitor the working environment are in various stages of development. Especially in Finland, Denmark, Sweden and the UK such a development is well under way. While in Germany and the Netherlands these developments are still in the preliminary stages. In the remaining countries, Austria, Belgium, Ireland and Spain, these developments are mainly limited to singular systems, e.g. in the field of occupational accidents or occupational diseases.

In Finland a comprehensive system of national registers and databases to follow the state and trends in safety and health is in operation. It covers: occupational injuries and diseases, industrial hygiene measurements, biological monitoring, exposure to carcinogenics, radiation exposure and chemical products. National surveys study work environment and workforce, and occupational data of national censuses have also been linked with health outcomes such as mortality, disability and cancer morbidity.The Finnish Institute of Occupational Health (FIOH) is developing its surveillance and coordination activities in order to improve the collection, production, refinement and dissemination of relevant data on the state and trends of the working life in Finland.

In the United Kingdom several new developments are reported:

- The Field Operations Division's new computer system FOCUS will improve the range and quality of data available to the whole of the Health and Safety Executive (HSE) and be of assistance in policy formulation as well as guiding inspection strategies
- HSE plans to simplify the reporting requirements and procedure of RIDDOR (Reporting of Injuries, Diseases and Dangerous Occurrences Regulations), and to incorporate the offshore and railway reporting arrangements.
- HSE plans a new strategy for obtaining improved information on the following areas:
  * frequency and causation of occupational diseases and injuries;
  * obtain an accurate estimate of the national incidence of occupational injury and its costs to the economy through the annual Labour Force survey;
  * make arrangements to obtain updated estimates of the national incidence of work-related illness and its cost to the economy through extended questions on the 1995 Labour Force survey;
  * develop proposals to allow employers the option of reporting by telephone to the RIDDOR-requirements.

In Sweden the Statistical Bureau of Sweden (SCB) compiles and analyses different matters by combining various sources of statistics. To coordinate linking and matching of data files there must be a special authorization in case the data files contain personal code numbers.

The Danish Working Environment Service states that with the existing knowledge and documentation it is possible to predict Denmark's health and safety problems in the coming years. An action program, announced by the government in 1994, is aimed at a continuous improvement of the working environment by preparing a midway report (in the year 2000) and a short status report for the Danish parliament every year.

In countries such as Germany and the Netherlands efforts are made to develop links between various data sources or totally new systems. In Germany over many years data has been gathered for use in preventive health care within the framework of the 'Humanisierungsprogramm Arbeit und Technik der Bundesregierung'. These projects run at various levels (companies, branches, regions). The accessibility of the data must however be improved and a clear structure should be developed to facilitate integration.

In the Netherlands, since 1993 the aim is towards an integrated monitoring system with basically three goals:
- to guarantee a basic level for the working environment
- to prevent (major) accidents/catastrophies
- to stimulate labour participation and to prevent fall out.

This monitoring system does not aim to be an information system but will have to function as a support system for the top management of the government organization by monitoring the effects of interventions on the working environment. Included in the system will be data on hazards/risk factors, intervention instruments and effects.

Furthermore in the Netherlands the topic of developing a reliable registration system for sickness absence data in cooperation between government and other agencies is currently being discussed. Finally, in the Netherlands the improvement of the occupational accidents statistics is currently being undertaken.

Also in countries with a more limited infrastructure concerning data gathering, such as Belgium and Ireland, some developments can be noticed. In Belgium on a governmental level, initiatives have been taken to classify data on occupational accidents according to the ISCO-system. The social partners however stress the need for a re-thinking of the strategy for data gathering. Their main point is not to accept without questioning the European claim for specific data if these data are not available in Belgium. It is more important in their view to consider alternatives to facilitate the process of providing existing data to European bodies. An important development mentioned furthermore is the establishment of a 'cross-point data bank' to link all available data on working conditions.

In cooperation between government and the National Association for the Prevention of Occupational Accidents (ANPAT/NVVA) in Belgium a 'working conditions barometer' is being developed. The aim is to present a representative view on working conditions in Belgian companies with the aid of this instrument.

In Austria occupational accident statistics, covered by the general Occupational Accidents Insurance Agency (Algemeine Unfallversicherungsanstalt, AUVA) and including data on occupations and sectors, will from 1995 carry classifications according to NACE. Furthermore, the AUVA conducts yearly evaluations of these data, e.g. for the Social Insurance Association in Austria and for other interested parties within or outside the AUVA.

In Spain there is an aim towards the development of information systems. This will be done through studies of existing statistical data, surveys, risk map methodology, sentinel health events, case identification by active search and prevalence studies. The approach of these systems depends on the objective of the action: surveys are mainly global, risk maps are sectorial and sentinel health events methodology is addressed to the totality of the working population.

Finally an interesting development was noted in Ireland. In 1993 a central database on occupational accidents was developed. Inspectors were issued in 1994 with portable computers which enable them to transfer inspection data between their portable computers and the central system database. Inspectors will be able to print out enforcement notices from the system and issue them on the spot. They can also access other databases through their portable computers so as to provide employers with advisory material.

6.2.2 Sectorial approach

From the data provided for this research is apparent that in 7 countries developments are apparent by way of a sectorial approach to data gathering in the field of working conditions. These countries are Belgium, Spain, the Netherlands, Ireland, the United Kingdom, Austria and Denmark. It is reasonable to assume that in other countries also certain forms of sectorial approaches are in operation, however this information was unavailable for this report.

In Belgium a study is developed on the marine fishing industry by the FAO (Fund for Occupational Accidents). Furthermore, within the framework of the Foundation's project on sector profiles (European Foundation, 1995b, 1995c), the intramural health care sector and meat processing industry have been studied, as has been done in 9 other EC-countries. In Belgium a specific follow-up to this project has been planned.

In Spain the construction sector has received special attention. At national level there is a Foundation with the main objective to improve the working conditions in the construction industry.

Furthermore women as a specific risk group are receiving more attention from the government as well as from the unions. In Catalunya there is a Construction Committee (CSHCC) with the same objectives mentioned above. One of the specific objectives of this committee is to develop a database and a bibliographic collection about working conditions. They publish informative notes and "Safety notes". Chemical industries employers in Spain will be starting a campaign to change the public opinion of chemical industries.

In the Netherlands 11 sectors (branches of industry) have been targeted by the labour inspection (Arbo'91). Essential to this approach is the active support from the social partners. The main goal was threefold:

- to enlarge the awareness for working environment related problems among employers and employees;
- to stimulate self-activity among employers and employees;
- to improve the working environment.

In this approach the following procedure has been applied:

- Analysis of safety-, health- and well-being risks in the branches and an inventory of the policies on working environment conducted by the social partners and the companies. Within this stage a strong substantial and financial involvement of the social partners is required. The outcome is a description of the working environment of the sector.
- Formulation of the strategy: proposals which must contribute to the reduction of working environment related problems within the sector, are taken up. Agreements are made with the social partners about the responsibility for the execution of the proposals, and with which priority this will be done. The outcome is an action program (strategy note).
- Execution of the proposals.
- Reporting and evaluation.

The Dutch Labour Inspectorate has divided the field of work into 4 risk categories. Risk category 1 are sectors with the highest working environment risks and category 4 the lowest. The Inspectorate has made this division on the basis of analysis of percentages of sickness absenteeism, on accidents, occupational diseases, occupational disablement and on prevalent hazards of the working environment. On grounds of priorities and starting-points an inspection program was planned for the period of four years ('91-'94). For this period the percentage of inspections for the various risk categories was planned dependent on the risk level. Furthermore companies with less than 10 employees will be inspected once every four years; companies with more than 10 employees twice every four years. This means that all branches of industries will be visited by the Labour inspection throughout the years.
The Inspections of the Ministry of Social Affairs and Employment in the Netherlands will be more structured in the future. Important changes will be the reduction of the number of items within the inspection list and the attention on 'health and safety management systems'. Furthermore the Labour Inspectorate will retreat and risk inventory and evaluation has become the sole responsibility of the companies. These companies are however obliged to use the professional advice of certified Occupational Health and Safety Services.

In the United Kingdom the government stated the following sectorial approaches:

- Improvements in the quality of offshore inspection work by ensuring that issues raised by the Safety Cases are followed up.
- On the railways, enforcement of health and safety becomes operational.
- Continue to improve the level of compliance with COSHH in sheep dipping operations through routine inspections visits.
- Keeping safety at operating sites as a top assessment priority.
- Increasing the number of inspectors dealing with construction to work more closely with the

industry to encourage the improvement of health amd safety management under the new Construction (Design and Management) Regulations.

In Ireland the sectorial approach led in 1993 to inspections by the Health and Safety Authority (HSA) in 12 areas:
- factories/manufacturing industries;
- construction;
- agriculture/forestry;
- mines/quarries;
- health care;
- education sector;
- offshore installations;
- fishing industry;
- retail outlets;
- hotels;
- public authorities;
- disciplined services;

Furthermore a number of activities and targets have been agreed for the forthcoming year by the various sectors with HSA. These are joint efforts with each of the industry sectors and HSA staff. Meetings, exhibitions, training, publications, publicity campaigns, use of the media are all part of the continuing campaign to improve awareness and raise standards within all industries. Of particular significance is the increasing reference to workplace health and safety matters in articles and interviews from sources outside the Authority.

In Austria a sectorial approach has been adopted by using the NACE-classification system in the occupational accidents and diseases statistics.

Within the action program in Denmark prevention objectives are indicated for specific sectors to eliminate health and safety problems.

6.3 Discussion and preliminary conclusions

It should be clear that in the previous paragraphs only a limited account could be given of the many problems and developments regarding health and safety issues and data gathering in the participating countries. The level and quality of information available to the researchers moreover is diverse in nature and it can be assumed that for several countries, e.g. Sweden, more information probably exists. However due to time constraints the research had to be concluded, which consequently means that for some countries the 'picture' might seem incomplete. Nevertheless the researchers feel confident that the aim of this chapter has been largely met, i.e. highlighting the main problems in working conditions and developments with regard to data gathering for each country. In the final paragraph of this chapter some conclusions are drawn on the basis of the previous paragraphs. Furthermore the discussion in this chapter will aim at positioning the research results described in the previous paragraphs into a European perspective.

In order to accomplish this a number of issues will have to be analysed. To facilitate this analysis a table is constructed in which very briefly the situation in each country is characterized.
The aim is to ascertain whether there is a relation between asserted main problems and the topics to take initiatives on and/or current activities and developments.
In this table 6.1 for each country that provided data in this respect 3 items are assessed in general:
- statistical problems, i.e. is it possible to obtain various statistics on exposure and health and

safety output.
- assessment problems, i.e. is it possible to obtain data on hazards, materials, substances, risk-groups etc.
- initiatives, current activities or developments.

Table 6.1: ***Overview of reported problems and activities per country***

Country	Statistical problems	Assessment problems	Topics/activities/developments
Austria	no	insufficient data	limited
Belgium	insufficient data	insufficient data	limited, working group
Denmark	no	no	operative monitoring system /action-program
Finland	no	no	operative monitoring system /action program
Germany	lack of structure	insufficient data	action-program (limited effect)
Ireland	insufficient data	in development	limited action program
The Netherlands	lack of sickness absence data	no	design of monitoring system /action program
Spain	insufficient data	insufficient data	local/region oriented initiatives
Sweden	no	no	operative monitoring system
The United Kingdom	limited	limited	operative monitoring system /action program

A general theme in the asserted problems is that in at least 7 of the 10 countries (that provided information on the issues at hand) no comprehensive system for monitoring the working environment and health and safety output is in operation. The general picture is (with exceptions) that necessary (statistical) information is lacking - or if available, is often very limited or unreliable to be able to structure the priorities and guide interventions in the working environment. Looking at topics to take initiatives on and current developments in the countries involved, a number of activities are in operation to solve these problems. In at least 7 countries initiatives are deployed to integrate existing systems or develop new monitoring systems.

Although it is not clearly established on the basis of this research, it can be assumed that comparable problems (or worse) can be found in the rest of the EC-countries. It is largely unknown to what extent in these countries which initiatives are being deployed to solve these problems.

Within the various countries involved in developing monitoring systems no trends in cooperation could be observed between countries. In other words there seems to be limited communication between countries on e.g. the best available methods of registration, or the most adequate indicators to use, or on a different level the best way of implementing strategies to improve working conditions.
The various initiatives seem to be limited to 'local' application only. It is clear that this will not facilitate easy access or easy comparability of data from the various Member States. Moreover the

production of 'European figures' will be further hampered by this diversification. At the European level however initiatives have already been deployed, e.g. Eurostat ESAW and EODS-projects (see section 1.1). This seems a successful approach to coordinate a uniform way of data gathering.

The information received further shows that in most of the countries the governments, and affiliated organizations or agencies, are the main actors in the field. However in formulating action-programs and implementing policies many governments are often in cooperation with specialized institutes, as e.g. in Denmark and Finland. In only a limited number of countries do social partners seem to try to influence policies in the field of working conditions, especially with regard to data gathering and production.

Within the framework of defining new risk factors or hazards and new risk groups it is especially interesting that in some countries the rise of new technology was mentioned as a risk factor. Furthermore there seems to be a process in operation of slightly shifting the focus from 'traditional' (physical) hazards towards more 'modern' hazards (stress).
In some countries it is the smaller and medium sized companies (less than 200 employees), where in general 50-60% of the workers are employed, which are mentioned as a group where it is often very difficult to obtain information on working conditions. Also the enforcement of regulations in this group is considered more difficult than in larger companies. It is likely that this enforcement consumes much of the resources of e.g. inspection bodies in the various countries.

7 *Final conclusions and general discussion*

Introduction

In this chapter, all previously presented data and information will be discussed from an overall perspective, which will result in the final conclusions. In this respect, this chapter forms the synthesis mainly of the sections 4.3, 5.3 and 6.3, which contained discussions and preliminary conclusions on specific subjects.

The overall perspective for the final discussion and conclusions in this chapter is in fact set by the main objectives of this pilot study: to furnish policy makers at European level, and at national level, with information which enables them to set priorities in measures to improve the working environment and the monitoring thereof. More specifically, the final discussion and conclusions should address the aims of this study, as they were described in section 1.2:

1. To describe the working conditions in Europe based on existing data and information sources.
2. To identify common exposures and risks among workers across Europe, especially in relation to gender, age, occupation and sector of economic activity.
3. To supplement the 'subjective data' from the First European Survey on the Working Environment 1991-1992 with 'objective' data from sixteen European countries.
4. To assess the extent of the lack of comparable data in Europe, and discuss the problems related to this.
5. To identify where further data gathering, quality improvement, or harmonization is most urgent.

As was already mentioned in chapter 2, due to major deficiencies in the availability and quality of the data obtained, the emphasis in this report has come to lay more on the aims 3, 4 and 5, regarding the quality of the monitoring, than on the aims 1 and 2, regarding the quality of the European working environment. This distinction will also be reflected in this chapter, inter alia by the sequence of the sections. Beforehand can already be mentioned that aims 1 and 2 could only be met to some extent, aim 3 could not be met, and aims 4 and 5 have well been met.

Section 7.1 presents the discussion and final conclusions on the monitoring of the working conditions and health and safety output at European level. It addresses aims 3, 4 and 5 and provides answers on the following questions:
- Which data are available and how reliable and comparable are they?
- From what kind of information sources do they stem, and what organisations are involved in data production?
- To what extent can the data be used to describe the European working environment?
- Which developments in monitoring are taking place?
- On what issues is further data gathering, quality improvement, or harmonization most urgent?

Section 7.2 subsequently addresses aims 1 and 2, by discussing and concluding on the outcomes of the monitoring efforts in the various countries. The following questions are central here:
- Which main risk factors in the work environment across Europe can be identified?
- Which main risk groups can be pointed out?
- What does the data say about the health and safety output among the workers in Europe?
- Can trends for the future be identified?

7.1 The availability and quality of data

To structure the discusssion and final conclusions regarding the monitoring of occupational health and safety at European level, the scheme in figure 2.1 will be used as a reference, which visualises the structure of a 'theoretically ideal' monitoring system (see section 2.1). 'Ideal' here mostly means that the system is comprehensive and maximally or optimally filled with data on the three monitoring areas defined (working environment, interventions, health and safety output). Yet, it should be stressed that what is 'ideal' is in fact a point of discussion, and very much related to the practical purposes of the system. These practical aspects will be dealt with in chapter 8, where recommendations for future monitoring will be given.

Hereafter, this concept of an ideal monitoring system will be confronted with the findings in this study on the actual state of the art on monitoring occupational health and safety in Europe. This leads to conclusions and discussion, both with respect to the actual monitoring system as a whole, as well as on the three monitoring areas separately. Besides going into the availability, reliability and comparability of data, information sources and trends and developments in data production will also be dealth with. In all sections, conclusions and points of discussion will be clearly distinguished.

7.1.1 Availability of data

To properly interprete the conclusions and discussions hereafter, it is firstly repeated what was mentioned at the introduction of chapter 4 regarding the meaning of the terms 'available' and 'unavailable'. 'Available' means that national reports actually provided data on a specific topic. 'Unavailable' means that national reports contained no information on that topic (blanks), or had the remark 'no data available'. This may have several backgrounds, however. Data may not exist at all, or data may exist, but not according to the structure of the matrix used for national data gathering, and/or not accessible or convertable within the time restrictions of the project. It means for example, that data on risk factors may exist in a country from various studies in branches, while the national report mentioned 'no data available', because data were not convertable to the entire national working population. Further analysis of the background of the 'unavailability' of data may reveal interesting additional information.

Availability in general
Firstly, it is concluded that on half of the topics covered in this study, sufficient data is available from national information sources to fill a European monitoring system. On fifteen of the topics six or more countries had data available. On some topics, also on the other fifteen, additional sources at European level are available however, which might provide sufficient data.

The second conclusion is that data is not equally available within the three monitoring areas. The focus in the various countries is more on the health and safety output, than on the working conditions themselves, whereas the interventions seem least being monitored.

This second conclusion is inter alia based on the observation that data on five out of ten health and safety output topics were provided by 9-15 countries, whereas data on eleven out of twenty risk factors in the working environment, stemming from national sources, were obtained from only 6-10 countries. This is more or less mirrored by the situation at European level: as section 1.1 has shown, cross-national studies and initiatives on harmonization of data more often concern health and safety output topics, than occupational exposure data. The First European Survey on the Working Environment (the 'Eurosurvey'; European Foundation, 1992) moreover stems from only

1991-1992, whereas initiatives from the European Commission to harmonize occupational health and safety statistics go back to 1962.

Lastly, it is concluded that only on the variable gender sufficient data is available on a number of topics in both the working environment, as well as health and safety output. Data broken down to age and economic sector is sufficiently available on only a limited number of topics (four health and safety output topics, respectively three in total). Data according to occupation is insuffiently available on all the items studied (data available in less than six countries).

Working environment
Firstly, from the information in chapter 4 it is concluded that to some extent, data is sufficiently available from national sources to describe the working environment at European level. Particularly (aspects of) the physical, psychosocial and ergonomical European working environment may be described, but almost exclusively based on workers' perceptions, and rarely on 'objective' measurements.

More specifically, sufficient data is available on the number and percentage of workers exposed to eleven risk factors: hazards related to noise, vibrations, climate, radiation, working pace, job content, working hours, influence and control over work, social interactions, working postures, and manual handling.

For a European monitoring system based on national sources, data is insufficiently available on nine other risk factors: workers exposed to hazards related to lighting, pressure, form of payment, traumatic experiences, handling of tools and equipment, chemical substances, biological agents, safety and other risks. Sufficient data is available however from the Eurosurvey on three of these risk factors: form of payment, handling of tools and equipment, and chemical substances.

It is further concluded that data is only sufficiently available on some risk factors for the variables economic sector (job content, working hours) and gender (noise, vibrations, climate, job content, working hours, social interaction, manual handling). On none of the risk factors studied, is data sufficiently available according to age and occupation.

Lastly, it is concluded that in six countries no working environment data at all are available from national sources. Information on the working environment in Greece, Ireland, Italy, Luxembourg, Portugal and the United Kingdom is only available from the Eurosurvey.

Interventions
Monitoring of interventions seems to be not common in the various countries. Hence, data from national sources on interventions to improve working conditions and reduce health consequences, are probably insufficiently available to fill the 'ideal' European monitoring system.

It should be noted however, that this study can in fact provide not much insight in this area, because it was not explicitly included in the matrix used for the national data gathering. At a global level information could be provided though, based on national policy documents and on information from key informants (see chapter 6).

In fact, only the Netherlands explicitly reported the development of an integral national monitoring system which will include interventions. Yet, national (statistical) data on topics such as health and safety committees, Occupational Health Services, workplace assessments and action plans are known to exist in several countries. However, at European level only one study on the subject of interventions is known, which was only recently completed (European Foundation for the Improvement of Living and Working Conditions, 1995d).

Health and safety output

Firstly it is concluded that, in order to fill the 'ideal' European monitoring system with health and safety output data, sufficient data is available from the national reports on five topics, of which three are directly work-related: occupational accidents, occupational diseases, and occupational mortality. The two more general topics regard sickness absence and general mortality.

On the five other topics studied, data is insufficiently available. Again three of these topics are directly work-related and two have a more general background: occupational sickness absence, occupational morbidity, and occupational disablement, respectively general morbidity and general disablement (see tables 4.1.1b and d).

It is further concluded, that for the topics for which sufficient data is available, data is mostly available broken down by gender and age. The only exception is occupational mortality for which less than five countries provided data. Data broken down by economic sector and occupation is rarely sufficiently available: of all ten topics studied only on occupational accidents, and only broken down to economic sector.

The last conclusion reads, that the availabilty of health and safety output data is rather limited in Norway, Portugal, Ireland and France, varying between none (N) to three topics (F). This finding seems odd and it might be induced by time restrictions of the project. From the twelve other countries, data is available on four (GER, GR, I, L) to eight (FIN) topics.

Relationship between causes, interventions and effects

The main conclusion on the inter-relatedness of the data, i.e. the ability to link causes, interventions and effects, is that there are clearly restrictions to this aspect of the 'ideal' monitoring system. Firstly, because data on interventions are overall hardly available (see above) and secondly, because data on health and safety output according to diagnoses are only limited available.

In fact, data on work related causes of health and safety output are only sufficiently available regarding occupational accidents and occupational diseases. Diagnostic data are also sufficiently available for general mortality, but these are not directly work related and can therefore only be indicative. Furthermore, the ability to link data on occupational accidents with working environment data is severely impaired by the fact that data on safety risk are insufficiently available at European level.

Discussion

The differences in extent of activity in the three monitoring areas, mentioned earlier under 'Availability in general', may be explained by historical factors. Awareness of bad working conditions usually starts when serious health problems among workers arise. Great numbers of (fatal) accidents at work or work related diseases alarm society and policy makers, who then start asking for insight to the extent of these problems. Subsequently, questions concerning the causes of these health problems arise, and insight to the extent of occurence of risk factors is then also required. Based on that information, policy actions are taken to reduce and prevent the problems, which after some time evokes the desire to get information on the effectiveness and efficieny of these interventions as well.

This historical development of national monitoring systems is further reflected by the socio-economic arrangements and provisions in countries, which may also explain why output data is generally more avaibable, than data on the other areas. Health and safety output is strongly related to general or occupational social security systems and obligatory registrations, and the fact that not only people's life or health are at stake, but that also enormous amounts of money are involved, is probably a strong incentive for the production of such data. Yet, objective data on the working environment in companies and institutes, and on the preventive and corrective actions undertaken, exist in fact as well in all countries. For example due to governmental inspection and enforcement activities, to work place assessments, and to specific research. However, these data are generally not aggregated to national or sectorial level in a standardized way, because other aims are strived after rather than national monitoring, let alone cross-national monitoring of the working environment and improvement activities. Yet, regarding the sectorial level, some developments can be observed (see section 7.1.5).

With respect to the historical development of national monitoring systems it should further be mentioned that the stage of development varies between the various countries. When combining the information on the availability of data in the various countries (section 4.1), with information on the developments towards comprehensive monitoring systems (section 6.2.1), three stages can roughly be identified. Comprehensive monitoring systems are established in Finland, Denmark, and Sweden. Developments in this direction seem well under way in Germany and the Netherlands, and to a more limited extent in the United Kingdom, Austria, France, Spain, Belgium and Norway. Main limitations in the United Kingdom and Norway are the coverage of only health and safety output, respectively the working environment (if this is a correct observation). In Austria and France the working environment data are almost ten years old, in Belgium they cover only a limited number of risk factors, and both in Belgium and Spain they stem from only one information source. The national monitoring systems of Greece, Italy, Luxembourg, Portugal and Ireland seem to be in the lowest stage of development, producing only limited data on health and safety output.

As another point of discussion, it should be noted that monitoring of output data rarely concerns all economic costs involved in occupational health and safety. Although this subject was not specifically covered in this study, because for several reasons it was considered to be too complex to incorporate it, this conclusion seems valid from the discussions in the Steering Group and from the national data obtained. From some countries it is known that data on the costs involved in the existing compensation arrangements on sickness absence, invalidity, occupational diseases etcetera, are available to some extent. But costs like investments in workplace improvement, in worker's training, in Occupational Health Services, or the costs of inspection and enforcement, other interventions, and of work related curative health care are usually unknown. This was quite dramatically demonstrated by a study in the United Kingdom, which for the first time tried to provide a picture on this matter (Health and Safety Executive, 1994). Consequently, also at European level a reliable overview of the economic impact of (bad) working conditions to society is lacking, so far. Yet, estimations indicate that 20,000 million ECU are paid each year in the EU to compensate for injuries and illnesses (European Foundation for the Improvement of Living and Working Conditions, 1994e). Furthermore, the subject is gaining more attention, considering for example the Foundation's initiative to develop a European economic motivation model (same reference).

Hence, although the observed focuses in occupational health and safety monitoring across Europe are explainable, and taking into account that the limitations in availabilty of data may be induced by the matrix used in this study for the national data gathering, the conclusion seems justified that necessary information to enable a European preventive policy towards health and safety at work, is insufficiently available when taking the 'ideal' monitoring system as a reference. Particularly the standardized monitoring of the working environment, the interventions, and the economic costs of work-related accidents and ill-health to society, could be intensified.

7.1.2 Reliability of data

In reference to the 'ideal' European monitoring system, the reliability of the data obtained is generally concluded to be good with respect to the validity, the representative value, and the up-to-date-ness. The objectivity or 'hardness' of the data is problematic, however. No conclusions can be drawn on the reproducability. All these aspects of the reliablity are specified hereafter.

Validity
The conclusion that the validity of the monitoring data is good, is not based on a thorough analysis of the methodologies used in the national information sources. It is based on the observation in section 4.1.2 that these sources stem from authoritative and professional organisations with long-standing experience, which are assumed to carry out their data gathering activities in accordance with scientific standards.

Uptodateness
The up-to-date-ness of the data is concluded to be good, because the health and safety output data generally stem from 1992-1993, whereas the working environment data mostly stem from either 1992-1993, or from 1990-1991 (see tables 4.2.2.a and b).

Representative value
The representative value is further considered to be in order, because on almost all sixteen topics on which data are sufficiently available, data from six countries or more cover the entire national working population. Only on radiation, working pace and general sickness absence less than six countries have nationally representative data (see tables 4.1.2.a and b).

Hard data
The reliability of the data obtained is generally concluded to be insufficient with respect to its objectivity or 'hardness'. For the working environment data, this is due to the subjectivity of almost all information: the data on all eleven risk factors for which data is sufficiently available, stem in all countries concerned from questionnaire-based surveys. Only the Netherlands, France and Belgium provided some objective information on noise, vibrations, climate and radiation, but these were mere estimations (NL), or only representatieve for selected, yet large, goups within the national working population (B, F) (see table 4.2.1a).

The health and safety output data generally stem from objective sources such as statistical registers, but by no means they can be considered hard data, i.e. clear facts. Of all five topics for which data is sufficiently available, only general mortality data seem hard. Yet, these data are not directly work-related, and therefore not meaningful as such, but only in relation to occupational health and safety data. The data on occupational accidents and occupational diseases can overall not be taken as hard facts, because in several countries underreporting seems to be a significant problem, of which the exact extent is unknown. Probably related to this underreporting, at least in part, data on occupational mortality are not hard facts either, seemingly except for Denmark. Obtained sickness absence data is rather diverse in nature, and only more in-depth analysis of the data gathering methodologies used, could reveal how hard these data are.

Reproducability
As was stated above, no conclusions are drawn with respect to this last aspect of reliability of data, yet some discussion remarks are made. Reproducability can either be taken as 'getting the same results when using the same methodology again', or as 'getting the same results from using different methodologies or information sources'. In the first meaning, the reproducability is assumed to be in order for the same reasons that were mentioned for the validity of the data. In

the second meaning, assessing the reproducability requires comparing the findings of this study with results of others. To some extent this has been done successfully with the Eurosurvey, with respect to working environment data (see section 7.2). Further analysis is desirable here however, as well as comparing the health and safety output data with similar data from other studies, such as those mentioned in section 1.1. This particularly concerns general and occupational sickness absence, occupational accidents, occupational diseases, and general and occupational disablement.

Discussion
As a point of discussion, a conclusion drawn in section 5.3.2 with respect to occupational diseases, is repeated here. The conclusion read that countries with a sophisticated infrastructure for data gathering (and defining of diseases), report significantly higher rates than countries lacking such an infrastructure. The reason for repeating this conclusion is, that it indicates that more sophisticated registration and monitoring systems ensure a higher level of reliability of national data, due to a reduction of the extent of underreporting. Moreover, this seems applicable to other topics as well, both in the working environment and the health and safety output. A third point of interest related to this conclusion, is that differences in the level of reliability of national data, influence the cross-national comparison of data: different reliability levels between countries might lead to false conclusions on the European working environment and health and safety output, if this bias is not taken into account. These remarks are put forward here, mainly to draw attention to them. No conclusions are drawn in this respect, since required in-depth analysis of the various methodologies of the registration and monitoring systems used in the countries studied, was not feasible within the framework of this project.

7.1.3 Comparability of data

Comparability in general
Together with the fact that the data obtained are not hard facts, the cross-national comparability of the data is concluded to be the weakest aspect, in reference to the 'ideal' European monitoring system. Both for the working environment data, as well as the health and safety output data this deficit is due to the different definitions, classification-categories and sub-items used in the data reporting from the various countries. The working environment data moreover suffer from differences in the years from which the data stem (see section 4.2.2, tables 4.2.2a and b).

As a result of this deficient cross-national comparability, the identification of common risks and risk groups across Europe is concluded to be severely hampered. The use of deviating age categories and sectorial categories in the various countries particularly impedes the identification of common risk groups, whereas the cross-national differences in definitions, sub-items and reference periods especially impair the identification of common risks. Quite some effort is required to convert the available data into cross-nationally comparable information, which was left to future analysis, considering the time restrictions of this study.

Working environment
The cross-national comparability of the working environment data is slightly hampered by the fact that for all eleven risk factors on which data are sufficiently available, the data in only less than six countries stem from the same two-year reference period (see table 4.2.2a) Yet, data on nine risk factors (all except radiation, working pace) can be compared among at least six countries within a four-year period (1990-1991 and 1992-1993).

The comparability suffers severely however, from the use of different definitions, classification-categories and sub-items of risk factors. Only on noise is directly comparable data available from national sources in more than six countries, although minor difference were observed here as well. On the other ten risk factors comparison of data requires more in-depth analysis and probably needs conversion of categories and subsequently of data.

Based on these findings, it is concluded that at the moment, the best comparable working environment data available at European level, is the information from the First European Survey on the Working Environment.

Health and safety output
As far as the reference period is concerned, the cross-national comparability of the health and safety output data is concluded to be good. On all five topics on which sufficient data are available, the data stem from 1992-1993 in six countries or more (see table 4.2.2b).

The use of definitions, classification-categories and sub-items deviating from the ones provided by the matrix, however severely affects the cross-national comparability of health and safety output data. Only on occupational accidents, and in fact only on fatal accidents direct comparable data are available from at least six countries, The data on 'all accidents' and 'accidents with work interruption' are much lesser directly comparable, due to registration differences (1 day of absence, or 3 days of absence). General mortality data are comparable as well (although not in all respects in accordance with the matrix), but these data are not directly work-related. On the other three topics on which data are available from at least six countries, the data are generally not in accordance with the structure of the matrix. Assessment of their possible comparability otherwise, conversion requires more in-depth analysis and probably conversion of categories and data, as to avoid false conclusions.

Discussion
Although no clear information is provided, nor conclusions are drawn, it should be noted again that cross-national comparability of data might be affected by differences in the reliability level of national data (see the discussion of section 7.1.2).

As another point of discussion is put forward, that the observed lack of cross-national comparability might be determined, at least in part, by the observation in section 6.3 that most of the countries have set up national monitoring systems, without taking great effort in attuning theirs with the systems in other countries. The systems are primarily designed to serve national purposes, and harmonisation initiatives to serve international purposes may evoke resistance, for example because breaks in national statistics may arise, or compensation criteria and schemes would have to be changed. Initiatives towards harmonisation are therefore probably sooner taken at European level, than at national level. Yet, the Nordic countries Sweden, Norway, Denmark, Finland form an exception in this respect, since their working environment surveys are rather similar. But even between these countries not all data on the working environment are comparable, because data are available on several sub-items to varying extent.

However, although the 'local' orientation in setting up national monitoring systems is understandable, it should be mentioned that the national organisations involved may also gain from trans-national exchange of knowledge and experiences with occupational health and safety monitoring. If not for European harmonisation purposes or the production of European figures, then at least for learning and efficiency purposes at national level, international communication seems profitable. For example to identify the best available methods of registration, or the most adequate indicators to use.

7.1.4 Information sources and organisations involved in data production

General conclusions

Based on the information in chapter 4 (section 4.1.2), it is concluded that in all sixteen countries except Portugal, national information sources exist which contain quantitative data on the working environment and/or on health and safety output. Most common information sources are surveys (labour force; working environment), statistics and reports (annual; research). The main providers of information regard national bureaux of statistics, governmental bodies, national research institutes, social security organisations, and occupational insurance funds.

Data on Portugal appeared to be available only from international information sources: the Eurosurvey and the labour statistics from the International Labour Office (ILO). These international sources contain data on a majority of the other countries as well, although this was not always reported in the national reports.

Furthermore, based on the overview presented in section 1.1. it is concluded that at European and at global level more information sources exist with quantitative data on the various countries studied. These sources regard inter alia cross-national comparative studies by national research institutes, and statistics and reports from the ILO, Eurostat, the World Health Organisation (WHO), the Organisation for Economic Development (OECD), and the International Agency for Research on Cancer (IARC). The data from these sources however mainly regard health and safety output, and not so much the working environment.

Working environment

The existence of national information sources containing data on the working environment at national level has been reported by ten countries. Generally they regard specific national questionnaire-based working environment surveys, either held separately, or attached to the Labour Force Survey as so-called trailer questionnaires. Only in Belgium and the Netherlands the Labour Force Survey itself provides information on aspects of the working environment. Reports on Periodical Occupational Health Examinations and specific researches form another source of information, but seemingly only in the Netherlands and Belgium.

National information sources with nationally aggregated data on the working environment are entirely lacking however in six countries: for Greece, Ireland, Italy, Luxembourg, Portugal and the United Kingdom, the Eurosurvey is the only source available. Furthermore, France only has national information sources concerning physical, chemical and biological exposures. In Belgium it is the same, except for the fact that also an information source exist with data on one topic of the psychosocial working environment (working hours). The national information sources in the other eight countries (A, DK, E, FIN, GER, N, NL, S) cover a wider range of risk factors in the working environment: eleven (GER) to seventeen (DK) of the twenty risk factors studied.

The main providers of information on the working environment at national level are national research institutes and national bureaux of statistics. National governments seem much less occupied with actual data production at national level, although they are in fact main data gathering organisations, because of their inspection and enforcement activities. As was mentioned before, these data generally are not aggregated in a standardized way to national or sectorial level however, which would make them useful for national and international monitoring purposes.

Health and safety output

National information sources with health and safety output data at national level have been identified in all countries, except in Portugal and Norway. It is likely however, that they exist here as well. Main sources are statistics (general; working environment; labour force; labour market), surveys (working environment; labour force), and reports (annual; research). Existence of registers and databases were reported only by Spain, Denmark and Finland.

The national and international information sources consulted for the national reports on Portugal, Ireland and France only contain data on one to three out of the ten health and safety output topics studied. In the other twelve countries that provided output data, the information sources cover more topics: four (GER, GR, I, L) to eight (FIN) of the ten topcis studied. The national report on Norway did not contain health and safety output data.

The main providers of health and safety output data are national bureaux of statistics and governmental bodies, followed by social security organisations, occupational insurance funds and national research institutes. It is concluded that particularly governmental bodies, social security organisations and insurance funds play a far more important role in this monitoring area, than in data production on the working environment.

7.1.5 Trends and developments in data production

General conclusions
The first conclusion, based on information in chapter 6, which is assumed to be incomplete, regards a trend observed in practically all ten countries that provided data on these subjects. The lack of comprehensive, reliable (statistical) information at national level, specifically on causes and effects of working conditions, is increasingly identified as a predominant problem. As causes for this deficiencies, key informants mention the lack of a clear central organisational structure for national data gathering, insufficient access to data which does exist, and massive under reporting due to the limited scope of, or recent changes in the legal arrangements that form the basis for data production at national level.

The second conclusion is that particularly the national governments consider the inadequacies in information a main problem. They recognise the limited ability for setting national priorities and guiding interventions in the working environment. The social partners seem not so much concerned by this, and only in a limited number of countries they seem to try to influence policies in data gathering and production.

Another trend is, that the small and medium sized enterprises (SMEs; less than 200 workers) are increasingly recognised as a group which needs specific attention, also in data gathering. Although this was explicitly refered to by unspecified sources from only three countries (B, GER, UK), it is observed in other countries as well. The lack of an adequate organisation of data gathering is again mentioned as a causative factor.

Next, it is concluded that in all ten countries that provided data, developments in (the improvement of) data production are taking place, which are diverse in nature and extent. In seven countries (B, DK, E, FIN, GER, NL, S) initiatives are deployed on both the working environment monitoring, as well as health and safety output monitoring. These initiatives concern inter alia the improvement of already established integrated monitoring systems (DK, FIN, S), developments towards integration (B, GER, NL), or the improvement of singular monitoring systems (B, E, NL). In the other three countries (A, IRL, UK) the developments solely concern health and safety output. Conclusions on the most interesting developments (e.g. good practices) are given below.

The last general conclusion is that in eight countries developments inter alia concern sectorial data production. In five of them (A, DK, GER, IRL, NL), the developments in data production include an overall sectorial approach to monitoring which, except for Austria and Germany, is linked to sectorial programmes of governmental inspection and enforcement. From the three other countries (B, E, UK) it is known that specific sectors get special attention, either in research studies or in governmental inspection, which will contribute to sectorial data production on singular sectors. Although not explicitly reported, this is likely to be the case in other countries as well. In this respect is also refered to the sectorial studies mentioned in section 1.1, which were initiated at European level in 1993 and covered two sectors in ten European countries.

Working environment
With respect to the monitoring of the working environment, it is concluded that particularly interesting developments are taking place in three countries (B, DK, NL). In Denmark the interim results of a national action programme to improve the working environment, will be presented annually in parliament until 2005. In Belgium a 'working conditions barometer' is being developed by ANPAT/NVVA which will result in a representative overview of working conditions in Belgian companies. Since the late eighties, the governmental inspection of companies in the Netherlands has been sectorially organised, standardised and extended to the field of companies' policies (i.e. interventions). Data production at sectorial level has gained much by this development, because on all inspected sectors reports are published with aggregated data.

Other developments regard linking of data (S), the development of information systems (E) and further development of surveillance and coordination activities in data collection, production, refinement and dissemination (FIN). In Germany the aim is to facilitate integration of existing data by improvement of the accessibility and structure of data.

Health and safety output
Particularly interesting developments in the monitoring of health and safety output, are observed in three countries as well (IRL, NL, UK). In Ireland, labour inspectors were issued in 1994 with portable computers by which they can transfer inspection data to a central database on occupational accidents, which was established in 1993. In the United Kingdom the reporting procedure of injuries, diseases and dangerous occurences to the RIDDOR-system is planned to be simplified (e.g. allow employers reporting by telephone). Moreover, a strategy was planned by the Health and Safety Executive to update and extend the data on the economic costs to society of occupational injuries and work-related illness, via the 1995 Labour Force Survey. In the Netherlands the development of a reliable registration system for sickness absence data in co-operation between government and social security organisations, is currently being discussed. This has become a topic after a major change in the Dutch Sickness Act, which unintentially resulted in significant underreporting.

Other developments regard the improvement of occupational accidents statistics (NL), and converting them into occupation categories (A, B: ISCO) and sectorial categories (A: NACE). Furthermore, the developments regard the improvement of the range and quality of data in general by the introduction of a new computer system, and specifically of the frequency and causation of occupational injuries and diseases (UK). In four other countries, developments are similar to those mentioned for the working environment data production.

Discussion
When referring back to the three stages of development in which the national monitoring systems were roughly divided (see section 7.1.1), the information above on the developments taking place in the various countries, provides an additional perspective on the national monitoring. The developments in seven countries (B, DK, E, FIN, GER, NL, S) seem to match the stage of

development of the national monitoring systems. In the other three countries (A, IRL, UK) the developments in a way maintain a status quo in the stage of development, because they only improve health and safety output data, and not working environment data which in these countries are in fact deficient, either in availability or reliability.

To conclude this section on developments in data production, is once again refered to current initiatives at European level: the two harmonisation projects undertaken by Eurostat on occupational accidents (ESAW-project) and on occupational diseases (EODS-project) (see section 1.1).

7.1.6 Final conclusions

Positive points in occupational health and safety monitoring across Europe
Based on the previous discussion and conclusions, the following positive points in European monitoring can be summarized, when taking the 'ideal' monitoring system in Figure 7.1. as reference:

- on half of the topics studied with respect to the working environment, as well as the health and safety output, sufficient data is available from national sources to fill the monitoring system;
- data on the working environment and health and safety output generally seem valid, are representative for the entire national working populations, and are up-to-date;
- data generally stem from a variety of national information sources and the main providers of information concern five types of professional, authoritative organisations with long-standing experience in data production;
- interest in the monitoring of the working environment, interventions and economic costs is increasing, both at European level, as well as in various countries;
- interesting developments towards comprehensive, integrated monitoring systems are taking place in six countries and in six, partly different, countries initiatives are deployed to improve singular monitoring systems;
- several initiatives already have been taken to cross-nationally harmonize monitoring systems and statistics, both at European and national level.

Points for improvement in European occupational health and safety monitoring
From the previous discussion and conclusions, it is evident that the occupational health and safety monitoring at European level shows the following deficits, when taking the 'ideal' monitoring system as reference:

- unavailability of data, particularly concerning the working environment, interventions and economic costs, and to a lesser extent concerning health and safety output;
- unreliability of data, i.e. lack of objective data regarding the working environment, and underreporting concerning health and safety output;
- cross-national incomparability of data, i.e. differences in used definitions and classification-systems, and differences in sub-items of risk factors and of output issues on which data are available.

As a consequence of these deficits, the present 'European monitoring system' suffers from:

- limited ability to reliably pinpoint common or main risks and risk factors across Europe;
- limited ability to identify common risk groups across Europe according to gender and economic sector, and inability to identify risk groups among age groups and occupational groups;
- very limited ability to link causes, interventions and effects.

Furthermore, data production on small and medium sized enterprises calls for specific attention.

To conclude with, it may be considered problematic that particularly national governments regard the inadequacies in monitoring information a main problem, whereas social partners seem not so much concerned by this.

Final conclusions in relation to the aims 3, 4 and 5 of the study
Firstly, it is concluded that the third aim of this pilot study 'To supplement the 'subjective data' from the Eurosurvey with 'objective data' from sixteen European countries' can not be met with the material obtained. This is because objective working environment information, aggregated at national level, appeared to be too scarce. Furthermore, there are not really hard data on occupational health and safety output.

With respect to aim 4 of the study 'To assess the extent of the lack of comparable data in Europe, and discuss the problems related to this', it is concluded that this objective has successfully been met. The previous sections in this chapter have identified in detail which topics suffer from lack of data, from unreliability of data, and from cross-national incomparability of data (provided the matrix which was used in the data gathering, and in reference to the 'theoretically ideal European monitoring system'). Furthermore, the influential factors behind the unavailability, unreliability and incomparability of the data, have been pinpointed and discussed.

Finally, it is concluded that aim 5 'To identify where further data gathering, quality improvement or harmonization is most urgent' can also be met with this study. That is to say, the topics for which further data gathering, quality improvement or harmonization are indicated, can be identified. Simply put, this would mean the following:

- further data gathering is indicated on all topics on which data appeared insufficiently available: nine topics of the working environment; interventions; five topics of health and safety output, plus economic costs;
- otherwise making data available is indicated with respect to the working envrionment, i.e. aggregation of existing objective data, for example from national inspection and enforcement activities;
- improvement of the quality of data is indicated for all the health and safety output topics which suffer from underreporting (at least three);
- data harmonisation is indicated primarily for ten working environment items, two health and safety output items, and for age categories and occupational categories.

Chapter 8 will go into these areas for improvement in further detail.

Yet, the determination of 'what is most urgent' greatly depends on the policy aims that should be met with an improved European monitoring system, and the practical boundaries which set the conditions to the efforts and investments in this improvement. It is concluded therefore, that determining which improvements in the present European occupational health and safety monitoring are most urgent, requires a discussion at European level, involving policy makers and experts concerned with data production. Chapter 8 will elaborate on this discussion as well.

7.2 The working environment and health and safety output

Working environment
In refering back to the objectives 1 and 2 of this study (see introduction of this chapter), it is firstly concluded that the working conditions in Europe can be described to some extent. Based on the obtained data from existing information sources, descriptions can be made on aspects of the physical, psychosocial and ergonomical working environment in Europe, from workers' perceptions, as well as on safety at work, from occupational accidents registrations. Descriptions of the

chemical and biological aspects of the European working environment can not be made on national data, due to lack of sufficient data. However, data from the Eurosurvey allow a description of chemical aspects of European working life, to some extent. This means that only the biological aspects of the European working environment can not be described.

The second conclusion is, that there are indications that four risk factors are most common across Europe. These indications are based on both the obtained data from the national sources in ten countries, as well as the results from the First European Survey on the Working Environment (European Foundation, 1992) held in twelve, partly different countries (see section 5.1.3 and tables 5.1.1 - 5.1.3). The four most common risk factors are:
- defective job content;
- working pace;
- lack of influence and control over one's own work;
- strenuous working postures and movements.

These risk factors are considered 'most common', since exposure to these hazards is reported over the largest (extrapolated) parts of national working populations across Europe.

It should be stressed, that these indeed are mere indications and not hard facts. On the one hand, because they are based exclusively on subjective data, which are not confirmed by objective data from for example workplace assessments, or environmental measurements. On the other hand, because of differences in the methodologies used to compile the figures, which hamper comparability. Further analysis of the data obtained could perhaps provide more insight in the 'hardness' of these indications.

Thirdly, it is concluded that the other sixteen risk factors studied can not be positioned in an overall ranking across Europe. This is either due to lack of sufficient, comparable data, or to inconsistency between the data from the national sources, and the Eurosurvey (see section 5.1.3.).

No conclusions are drawn with respect to the severity of exposures, because no sufficient data was available on exposure levels, nor duration. However, further analysis of national data from five countries (DK, E, FIN, N, S), and data from the Eurosurvey could perhaps provide indications for the exposure severity of some of the risk factors studied.

Health and safety output
Identification of common risks across Europe is concluded to be possible, based on the data obtained on three health and safety output topics, but again this should only be taken as indicative, and not as hard facts (see section 5.3.2). These indicative common risks are:
- main causes of occupational accidents:
    * overload/extension of the body;
    * slips, trips and falls;
- main diagnoses of occupational diseases:
    * skin diseases;
    * musculo-skeletal disorders;
    * hearing disorders;
- main diagnoses of general mortality:
    * diseases of the circulatory system;
    * neoplasms.

This identification is only indicative, because the data are subject to significant underreporting, and to major cross-national differences in the methodologies used to compile the data (occupational accidents and diseases), and because they are not directly related to the work environment (general mortality).

Based on the other seven health and safety output topics studied, no conclusions can be drawn in relation to common risks. This is mostly due to lack of sufficient, comparable data.

Main risks
In the indications for common risks in both the working environment data and the health and safety output data, overlap is found on one risk factor. It could therefore be concluded that 'strenuous working postures' and 'overload/extension of the body' form main hazards for workers across Europe, often resulting in musculo-skeletal disorders.

The other eight indicative common risks are not so directly inter-related in terms of causes and effects. Identification of other main risks across Europe from the now available data, is therefore not justified.

Risk groups
To some extent, risk groups across Europe can indicatively be identified from the data obtained on the working environment and the health and safety output. They regard male workers, workers in some economic sectors, and workers in some countries. The reasons for the identification again being only indicative, are similar to those mentioned above.

Compared to female workers, male workers have overall higher rates of people exposed to noise and vibrations, as well as higher rates of accidents with work interruption and fatal accidents (higer rates of fatal cases particularly in Denmark, Spain, Germany and Italy).

The (metal)manufacturing industries and building/engineering sectors are concluded to be relatively hazardous sectors across Europe in relation to occupational accidents. In addition, agriculture, energy and water, and transport face relatively high rates of fatal accidents in Italy and Spain.

People in Denmark, Sweden, Belgium and Austria have higher rates in relation to general mortality, than people in other European countries. Whether and to what extent this is induced by work-related factors has not been analysed.

No risk groups among age groups and occupational groups could be identified, neither from the working environment data, nor from the health and safety output data, which is due to lack of sufficient, comparable data.

Trends for the future
The conclusion on trends in risks is not based on an analysis of the quantitative data obtained, but on interviews with key informants in ten countries (see section 6.3). In some countries, the rise of new technology is considered a risk factor. Furthermore, key informants report a shift in focus from the 'traditional' physical hazards towards more 'modern' hazards, such as stress.

8 *Recommendations for future policy*

Based on the conclusions presented previously, this final chapter provides recommendations in order to fullfil the main purpose of this report: to furnish policy makers, primarily at European level, but also at national level, with vital information which will enable them to assess whether the topics in today's occupational health and safety policies are still current, to determine which other improvement actions or interventions are necessary or desirable, and to set priorities therein. The recommendations will include both the working conditions, as well as the monitoring thereof. By giving these recommendations, the cycle in this study, which was schematically presented in figure 2.2 in chapter 2, will be completed.

8.1 Recommendations on the working conditions across Europe

Since in section 7.2 it was concluded that the obtained data justifies no reliable pinpointing of common risk factors, common risks or common risk groups across Europe, therefore the recommendations on these subjects can only be indicative. The data obtained indicates that priority in European prevention and improvement strategies should be given to actions targetted at the main and common risk factors, which were identified as:
- strenuous working postures and movements, overload and extension of the body, musculo-skeletal disorders;
- psychosocial hazards, i.e. insufficient job content, strenuous working pace, lack of influence and control over one's own work, and other hazards related to stress;
- the hazard of slips, trips and falls;
- noise, being the work-related causative factor for hearing impairment;
- work-related causative agents for skin diseases, diseases of the circulatory system, and neoplasms;
- the future hazards related to new technology.

Priority in actions is also indicated with regard to the following common risk groups:
- male workers with respect to their exposure to noise and vibrations, and their proneness to occupational accidents;
- workers in the (metal)manufacturing industries, and the building and engineering sectors in relation to occupational acccidents.

The prevention and improvement directed actions could subsequently consist of:
- Literature-based in-depth analyses of the extent and severity of the problems related to the hazards, as well as inventories of the 'best practices' applied to prevent and reduce these hazards across Europe.
- Confronting the identified problems and the applied best practices with the current European occupational health and safety policy as formulated in the EU-directives, in order to determine which additional actions are necessary and which kind of interventions promise to be successful (formulating new policy).
- European campaigns or action programmes on the chosen policy topics to stimulate the undertaking of actual improvement actions at national, sectorial and company level. These actions may comprise: dissemination of practical information materials; adequate basic training and qualification of (future) workers; sharing of solutions, e.g. via Internet, solution databases, workshops, seminars and short courses; developing and experimenting with new designs of work places, equipment and work organisation; financial support.
- Evaluation of the campaigns and their results.

To ensure the successful implementation of these improvement actions, it is recommended to involve governments, employers' representatives, unions' representatives, health and safety experts, and other relevant experts, both at European level and at national level. Furthermore, it is recommended that the actions are in accordance with the current developments in the national working populations, such as ageing of workers and increase of women's participation on the labour market.

8.2 Recommendations on occupational health and safety monitoring across Europe

Taking into account some European figures mentioned in this study, such as 8000 reported fatal accidents, 4.5 million reported accidents with work interruption, and an estimated 20,000 million ECU paid each year to compensate for occupational injuries and illnesses, it is considered worthwhile not only to take actions to improve the working environment across Europe, but also to improve occupational health and safety monitoring at European level, to be able to more reliably set the necessary priorities in the working environment aimed actions. Suggestions and recommendations in this respect are presented hereafter in four sections: suggestions that simply flow from the observed deficits in availability and quality of data; recommendations for European level policy makers, including long term and short term strategies; and finally recommendations aimed at national policy makers regarding national activities.

Suggestions directly flowing from the deficits in data
The most obvious recommendations to improve the occupational health and safety monitoring in Europe would simply be repeating the final conclusion on aim 5 as described in section 7.1.6: make available data which proved to be unavailable, improve the quality of data which is unreliable, and harmonise data which is cross-nationally incomparable.

More specifically this would mean:

- Take actions to stimulate the production of data at European level on the 14 topics for which data is insufficiently available from national sources: exposure to lighting; pressure; form of payment; traumatic experiences; handling of tools and equipment; chemical substances; biological agents; safety and other risks; occupational and general morbidity; occupational sickness absence; occupational and general disablement.
- Take initiatives to stimulate the national and European aggregation of existing objective data (e.g. from inspection and enforcement activities) on the 11 topics for which only subjective data is sufficiently available from national sources: exposure to noise; vibrations; climate; radiation; working pace; job content; working hours; influence and control over work; social interaction; working postures and movements; manual handling.
- Stimulate the production of data at European level on interventions and economic costs involved in occupational health and safety.
- Deploy activities to improve the reliability of data on the three health and safety output topics on which data are sufficiently available across Europe, but suffer from underreporting: occupational accidents, occupational diseases, and occupational mortality. Sickness absence may fall in this category as well, as in-depth analysis of the data obtained may show.
- Take initiatives to harmonise data collection and reporting level across Europe, regarding the 12 topics for which data is sufficiently available, but not directly cross-nationally comparable, and for which harmonisation initiatives have not already been undertaken at EU-level: vibrations; climate; radiation; working pace; job content; working hours; influence and control over work; social interaction; working postures and movements; manual handling; general sickness absence; and occupational mortality.

- Ensure that the new and improved data production includes data broken down by age, gender, occupation and economic sector, according to predescribed and cross-nationally accepted and used categories (e.g. ISCO, NACE), and which will lead to data which is reliable (e.g. valid, up-to-date, representative, objective or otherwise 'hard', and reproducable).

However, this conclusion stemmed from comparing the findings in this study on the actual state of the art on occupational health and safety monitoring across Europe, with the 'theoretically ideal' monitoring system, as described in section 2.1. As already mentioned at the beginning of section 7.1, the 'ideal' European monitoring system is however a point of discussion, very much related to its practical purposes. Considering this, the improvement activities mentioned above are not put forward here as the ultimate recommendations from this study, because they seem not entirely practical. If only it were for the effort, time and money it would take to realise this 'ideal monitoring system'. These recommendations show a direction for future policy however, and can be taken as input for a discussion on the desirable and feasible European monitoring system.

Long term European strategy
The critical remarks above lead in fact to a main recommendation with respect to occupational health and safety monitoring across Europe, which has a long term perspective. It is advised that policy makers at European level intensify the discussion on what should be the most desirable and practically feasible monitoring system for Europe.

It is further suggested to take the following questions and points as additional ingredients for this discussion, besides the actual findings in this study and other relevant sources:

- Does 'Europe' want maximal, optimal or minimal steering influence on occupational health and safety and its monitoring in the EU-Member States?
- Which purposes would a European monitoring system have to fullfil, e.g. producing scientifically reliable figures, or facilitating scientifically justified priority setting for preventive and improvement actions, or is just global priority setting sufficient?
- Related hereto then, does 'Europe' want:
    * a maximal monitoring system (filled with reliable and comparable data on all topics, i.e. the theoretically ideal monitoring system)?
    * an optimal monitoring system (filled with reliable and comparable data on the most relevant topics in the working environment, interventions and health and safety output, which should be defined)?
    * a minimal monitoring system (filled with nationally existing or improved data, only on the topics which bear great socio-economic impact on society, probably being only health and safety output topics and thus limited in its capacity to facilitate prevention aimed policy making)?
    * no monitoring system, but making use of the existing national and cross-national data as they are currently, perhaps by having sectorial profiles developed of the ten, assumingly most risky sectors across Europe?
    * or a growing scenario, starting from the present state or a minimal monitoring system, and gradually developing it along clear marks in time to an optimal or maximal system?
- How can the matter of data production on small and medium sized enterprises adequately be addressed?
- Should social partners be more concerned about the inadequacies in occupational health and safety information across Europe, and how could this be improved? What can be their role in improving the monitoring in Europe?
- What needs to be known before preventive and improvement actions on working conditi-

- What needs to be known before preventive and improvement actions on working conditions could be formulated? The answer to this question should be balanced by the amount of effort it would take, both at European level and in the various countries involved, to get comprehensive, reliable and comparable figures.
- Since the improvement of the European monitoring system depends greatly on national contributions to it, national commitments to a European strategy on data production seems a pre-requisit. This not only calls for involvement of national data producing institutions in the strategy development, but also for the creation of win/win situations in this strategy, by which aims and interests at European and national level should equally be met. It is necessary to avoid imposing European decisions onto national institutions as much as possible, whereas European support to national and cross-national monitoring improvement initiatives should be provided as much as possible.

The recommended discussion could be facilitated by the following concrete activities at European level:

- Initiating a so-called 'scenario study' in which the five options for a European monitoring system as mentioned above (or other options) are fully worked out and studied regarding their consequences, advantages and disadvantages, inter alia by interviewing involved actors, both at European and national level. Such a study would provide supportive information for better decision making and strategy development regarding European monitoring.
- Assigning a working group of policy makers and experts on data production, both from European and national level, to carry on the recommended discussion, and to prepare a strategy proposal to the European Commission on the development of a European monitoring system, for example by the year 2000 and 2005. Making use of good monitoring practices in various countries could be given as a condition for the strategy proposal.
- The formation of a supportive network of institutions involved with data production at European and national level could be very helpful to this working group, for example to comment on, or work out specific aspects of the possible strategy. Considering the tremendous amount of potentially valuable, objective data produced on the working environment and undertaken interventions across Europe, such a specific aspect could for example be how to make information from national inspection and enforcement activities accessible and usable for a European (and national) monitoring system.

  This network is likely to consist of expert representatives from governments, centres of statistics, research institutes, social security organisations, and occupational insurance funds, since in this study these institutions were identified to be the main providers of occupational health and safety information. For specific aspects other experts could be included as well, e.g. representatives from the Senior Labour Inspectors Committee (SLIC) in the case of the example given above.

Short term European strategy

In addition to the long term strategy, recommendations are also given for a short term European strategy, which could be set to work parallel to the long term activities in order to realise motivating successes in a measurable period of time. Should the indicated long term perspective of developing a European monitoring system not be initialised at all, these short term recommendations may result in improvements in monitoring occupational health and safety across Europe although in that case the crucial question of which exact European policy purposes should be served by the monitoring activities, will not be addressed.

Monitoring in general

Regarding the occupational health and safety monitoring in general, four recommendations are given:

- As in the long term strategy proposal, for the short term strategy it is also recommended to consider the compiling of more European sectorial profiles, e.g. of the ten presumed most risky sectors across Europe. The two existing profiles on the Meat Processing Industry and the Hospital Sector have proved to be successful in identifying targets for European improvement activities (European Foundation for the Improvement of Living and Working Conditions, 1995b and 1995c). These profiles could supplement the indicative identification of common risk factors, risks and risk groups across Europe, in the present study. A strong argument for such a sectorial approach to problem identification, is that the sectorial level is increasingly being recognised as a right level for preventive actions to improve working conditions, hence the right level not only for problem identification, but for problem solving as well.
- A general procedure for assessing the European priorities in working conditions improvement actions could be applied. This procedure could consist of firstly carrying out studies (sectorial profiles or not), which allow rough problem identification. These could then be followed by more in-depth studies on the prioritised risks, risk groups etcetera, in order to assess the prioritised and best interventions. The procedure could be concluded by determining the way of implementing these interventions.
- Another general recommendation regards formation or assigning of working groups on monitoring topics which have been selected for improvement with respect to the availability, quality or comparability of data across Europe. These working groups could consist of expert representatives from the main providers of occupational health and safety data across Europe, as were mentioned with respect to the supportive network of institutions, under the long term strategy.
- Furthermore, it is recommended to have the further analyses carried out on the material obtained in this study, but was left to the future because of time restrictions. These more in-depth analyses might result in extra information from which additional policy priorities could be identified, either regarding the working conditions themselves, or the monitoring thereof.

Monitoring the working environment

Specifically with respect to the monitoring of working environment data the following recommendations for actions at European level are formulated:

- The European Foundation could up-date and improve the Eurosurvey, for example by increasing its frequency to once every 2 years, by enlarging the number of countries covered, by taking up more topics in the questionnaire (perhaps also including health and safety output topics), and by forming population samples which can produce representative data according to age, gender, occupation and economic sector. This improving and up-dating seems the best option at the moment, to get comparable data at relatively short notice and with a reasonable amount of effort.
- From the up-date of the Eurosurvey, which is planned for 1996, a rough problem identification can be drawn, which could be followed by more in-depth studies in order to choose the necessary and best interventions from a European level, targetted at risks, risk factors, risk groups or economic sectors.
- The production of reliable and cross-national comparable data on the assumed future risks, related to new technology and stress, should actively be stimulated and supported.

Monitoring the interventions
Regarding the monitoring of interventions the following action from European level might be considered:

- The establishment of a European database on good practices: a collection of practical success stories on innovative, inspiring policy interventions to prevent, reduce and control occupational health and safety problems. This database could perhaps not so much serve as a reliable source of quantitative information on interventions undertaken across Europe, but more as a qualititative source of inspiration for policy makers at European, national, sectorial and company level. The exact form of the database (paper or electronic), the organisation of the input and output, and how to stimulate the filling and use of the database could be elaborated by a working group, which could perhaps benefit from the experiences of the development of a technical solution database which is likely to start at short notice (Netherlands Institute for the Working Environment, 1995).

Monitoring health and safety output
With respect to the monitoring of health and safety output data the following recommendations are put at European level:

- Consideration could be given to the question whether further harmonisation initiatives should be prioritised. This question regards the following topics: sickness absence (general and occupational), occupational morbidity, occupational mortality, and disablement (general and occupational). In order to answer this question it is recommended to start with assessing the feasibility of harmonisation efforts by a more in-depth analysis of the cross-national comparative studies on these topics which have been undertaken in the recent past (e.g. those mentioned in section 1.1).
- Particularly if further harmonisation initiatives show to be unfeasible, consideration may be given to the setting up of a cross-nationally standardised trailer questionnaire to the Labour Force Survey in each European country, covering health and safety output topics. Although the information will be subjective, it may prove to be a valuable supplement to the existing objective, but unsubstantiated data, by which for example, the extent of under reporting may be estimated. Experiences from countries which already use similar trailer questionnaires could be utilised in the design and implementation of such a European questionnaire.
- A standardised method could be developed and tested to estimate all economic costs involved in occupational health and safety in the various European countries. The experiences in the United Kingdom, where such an estimation has been made quite recently, could be used as a starting point for this development.

National initiatives
The final set of recommendations regard actions which national policy makers and data producing institutions could undertake, independent whether there will be activities at European level or not (although attuning is recommendable, if and where applicable). Much of the suggestions in fact mirror those at European level:

- Carry on a discussion on the desirable and feasible national monitoring system, in relation to the policy purposes it should serve and, if considered relevant, in relation to the stage of development of the national monitoring system compared to the systems in other countries. This discussion could take place in a working group of data production experts and policy makers, i.e. with representatives from the government, centres of statistics, research institutes, social security organisations, occupational insurance funds, and other relevant organisations.
- Identify the topics on which data should be made available, reliable and perhaps cross-nationally comparable, and develop well defined action plans on how to do so.

- Particularly Greece, Ireland, Italy, Luxembourg, Portugal and the United Kingdom are recommended to consider whether quantitative working environment data at national level should be made more available, since such data does not appear to be available from national information sources. A similar consideration is recommended to Portugal, Norway, France and Ireland with respect to health and safety output data, if the observation is correct that such data is only available on a very limited number of topics.
- Participate in exchange of information on, or at least take notice of developments and good monitoring practices in other European countries, and of initiatives and developments at European level. This is in order to learn and to be able to improve the national monitoring system more effectively and efficiently, avoiding unnecessary time and money consuming investments. A practical form for this might be the setting up, or participation in already existing networks and working groups of data producing institutions across Europe. Regular meetings on specific monitoring topics, undertaking of methodological comparative studies, making agreements on cross-national exchange of data, and bilateral communication could be the core activities in these working groups and networks.
- Stimulate the European Commission and other relevant organisations at European level to support (cross-)national initiatives to improve occupational health and safety monitoring across Europe, for example by asking for facilities, initiatives, or financial support to the working groups' and networks' activities.

9 Bibliography

It should be noted that the bibliography presented here only includes the information sources mentioned in this consolidated report. It does not contain the great number of information sources of each national report which were used to compose this consolidated report. For these national sources is being referred to the national reports themselves (see the list at the end of the Bibliography). Annex 1 provides a list of organisations and contact persons where these national reports may be obtained.

Einerhand, M.G.K., Knol, G., Prins, R. et al (1995) *Sickness and invalidity arrangements - Facts and Figures from six European countries*, Ministry of Social Affairs and Employment, the Hague, the Netherlands. Published by VUGA Uitgeverij B.V. the Hague, the Netherlands. ISBN 90-5250-934-4.

European Commission (1962) Recommendation of the Commission 2188/62 to the Member States concerning the adoption of a European schedule for occupational diseases. *Official Journal of the European Community*, No L81 of 31.08.1962.

European Commission (1966) Recommendation of the Commission 66/462/EEC to the Member States concerning the adoption of a European schedule for occupational diseases. *Official Journal of the European Community*, No L147 of 09.08.1966.

European Commission (1990) Recommendation of the Commission of 22 May 1990 concerning the adoption of a European schedule for occupational diseases. *Official Journal of the European Community*, No L 160 of 26.06.1990, p.39.

European Commission (1993) *The availability of occupational exposure data in the European Community*, by M.H.P. Smith and D.C. Glass, Institute of Occupational Health of the University of Birmingham. Directorate-General for Employment, Industrial Relations and Social Affairs, Report EUR 14378 EN.

European Commission (1994) *Information Notice on Diagnosis of Occupational Diseases*, Public Health and Safety at Work Directorate (DG V), Luxembourg.

European Foundation for the Improvement of Living and Working Conditions (1988) *How Occupational Accidents and Diseases are Reported in the European Community*, Dublin, Report EF/87/68/EN, ISBN 92-825-7575-6.

European Foundation for the Improvement of Living and Working Conditions (1992) *First European Survey on the Working Environment 1991-1992*, Dublin, Report EF/92/11/EN, ISBN 92-826-4378-6.

European Foundation for the Improvement of Living and Working Conditions (1994a) *Monitoring the Work Environment - Report of Second European Conference in November 1992*, Dublin, Report EF/94/01/EN, ISBN 92-826-7315-4.

European Foundation for the Improvement of Living and Working Conditions (1994b) *Euro Review - Bulletin on Research in Health and Safety at Work*, Issue 1994-1 and 1995-1, Dublin, EF/94/29/EN/FR and EF/94/30/EN/FR, ISSN 1024-3240.

European Foundation for the Improvement of Living and Working Conditions (1994c) *Exposure Registers in Europe*, Dublin, Report EF/94/22/EN, ISBN 92-826-8737-6.

European Foundation for the Improvement of Living and Working Conditions (1994d) *Absenteeism in the European Union*, Dublin, Working paper No. WP/94/29/EN.

European Foundation for the Improvement of Living and Working Conditions (1994e) *Catalogue of Economic Incentive Systems for the Improvement of the Working Environment*, Dublin, Report EF/94/08/EN, Cat No: SY-82-94-876-EN-C, ISBN 92-826-2705-5.
Summary to this report: *Economic Incentives to Improve the Working Environment*, Dublin, Report EF/94/07/EN, Cat No: SY-82-94-884-EN-C, ISBN 92-826-7685-4.

European Foundation for the Improvement of Living and Working Conditions (1994f) *Monitoring of the occupational health and safety conditions in Europe - Review of general, national, regional and sectorial surveys on secondary data*, Dublin, Internal Working Paper.

European Foundation for the Improvement of Living and Working Conditions (1995a) *The European Health and Safety Database (HASTE) - Summaries of descriptions of systems for monitoring health and safety at work*, Dublin, Cat No: SY-85-94-810-EN-C (Report and floppy disk), ISBN 92-826-8856-9.

European Foundation for the Improvement of Living and Working Conditions (1995b) *Working Environment at sectorial level in Europe: The Meat Processing Industry - Consolidated report*, Dublin. In print.

European Foundation for the Improvement of Living and Working Conditions (1995c) *Working Environment at sectorial level in Europe: Hospital activities - Consolidated report*, Dublin. In print.

European Foundation for the Improvement of Living and Working Conditions (1995d) *Identification and Assessement of Occupational Health and Safety Strategies in Europe 1989-1994 - Consolidated report*, Dublin. In preparation.

Eurostat (1992a) *Labour Force Survey 1992 - print-out on specific topics selected for this study*, Eurostat, Luxembourg.

Eurostat (1992b) *Labour Force Survey: Methods and definitions 1992 series*, Eurostat, Luxemburg, Annex III, p. 35-36.

Eurostat (1994a) *European Statistics on Accidents at Work (ESAW) - Specifications for Phase I of Pilot Project on Work Accidents in Europe*, Eurostat E-3 (Working Conditions), Luxembourg, December 1994.

Eurostat (1994b) *European Statistics on Occupational Diseases (EODS) - Pilot Project Specifications for Cases recognised in 1995*, Eurostat E-3 (Working Conditions), Luxembourg, March 1994

Eurostat (1994c) *Enterprises in Europe, third report - Second preliminary version, Volumes I and II, July 1994*, European Commission DG XXIII and Eurosat, Luxembourg.

Health and Safety Executive (1994) *The Costs to the British economy of work accidents and work-related ill health*, by N. Davies and P. Teasdale. HSE Books, Sudbury, United Kingdom. ISBN 07176 0666X.

IARC (1992) *Atlas of Cancer Mortality in the European Economic Community*, edited by M. Smans, C.S. Muir, P. Boyle. Published by the International Agency for Research on Cancer, Lyon, France. IARC Scientific Publications No. 107.

ILO (1994a) *Code of Practice on the Recording and Notification of Occupational Accidents and diseases*, International Labour Office, Geneva, Switzerland.

ILO (1994b) *Year Book of Labour Statistics*, 53rd Issue, International Labour Office, Geneva, Switzerland.

Laursen, P. et al (1994) *Harmonized statistics of occupational diseases in the European Community - A proposal. Final report*, Danish Working Environment Service, Copenhagen, Denmark. Appendix C, p. 127-152.

Netherlands Institute for Preventive Health Care - TNO (1994) *Monitoring occupational health and safety in Europe: time constraints and its implications*, by S. Dhondt, Netherlands Institute for Preventive Health Care - TNO, Leiden, the Netherlands. Report No PG-94.033.

Netherlands Institute for the Working Environment, NIA (1995) *Personal communication*, Mr. H. Tönissen, Amsterdam.

OECD (1989 and 1990) *Employment Outlook*, Organisation for Economic Co-operation and development, Paris.

Prins, R. (1989) *Sickness Absence in Belgium, Germany and the Netherlands: a comparative study*, Netherlands Institute for the Working Environment NIA, Amsterdam, the Netherlands. Thesis.

Prins, R., Veerman, T.J., Andriessen, S. (1992) *Work incapacity in a cross-national perspective - A pilot study on arrangements and data in six countries*, Netherlands Institute fro the Working Environment NIA, Amsterdam, the Netherlands and Ministry of Social Affairs and Employment, The Hague, the Netherlands. Published by VUGA Uitgeverij B.V., the Hague, the Netherlands.

WHO (1994) *International Classification of Diseases ICD, version 10* , Geneva, Switzerland.

European Working Environment in Figures: national reports

European Foundation for the Improvement of Living and Working Conditions (1996). Dublin.

- *National Report for Austria*, report WP/96/07/EN;
- *National Report for Belgium*, report WP/96/08/EN;
- *National Report for Denmark*, report WP/96/09/EN;
- *National Report for Finland*, report WP/96/10/EN;
- *National Report for France*, report WP/96/11/EN;
- *National Report for Germany*, report WP/96/12/EN;
- *National Report for Greece*, report WP/96/13/EN;
- *National Report for Ireland*, report WP/96/14/EN;
- *National Report for Italy*, report WP/96/15/EN;
- *National Report for Luxemburg*, report WP/96/16/EN;
- *National Report for Netherlands*, report WP/96/17/EN;
- *National Report for Portugal*, report WP/96/18/EN;
- *National Report for Spain*, report WP/96/19/EN;
- *National Report for Sweden*, report WP/96/20/EN;
- *National Report for United Kingdom*, report WP/96/21/EN;

ANNEXES

1. Participating institutes and authors of national reports
2. Contents of the Matrix and guidelines for national data gathering
3. International Standard Classification of Occupation ISCO-88 (COM)
4. Statistical Classification of Economic Activity in the European Community NACE-1970
5. Exposure Classification System - a Proposal
6. Diagnostic Categories of Diseases
7. Overview on the total working population in 16 European countries
8. Age categories used in the matrix

Annex 1 Participating institutes and authors of national reports

The national reports on each country may be purchased from the European Foundation, but also from the following institutes and contact persons. These persons may also be contacted for more indepth information and names of institutes to contact in the countries they studied.

	Countries	Contact
-	Belgium Austria France Germany Luxembourg	Association Nationale pour la Prévention des Accidents du Travail, ANPAT/NVVA Brussels, Belgium Tel: +32 2 648 03 37 Fax: +32 2 648 68 67 Mr. Marc de Greef
-	Spain Italy Portugal Greece	Instituto National de Seguridad e Higiene en el Trabajo, INSHT Barcelona, Spain Tel: +34 3 280 01 02 Fax: +34 3 280 36 42 Mrs. Maria Dolores Solé
-	Denmark Finland Sweden Norway	Danish Working Environment Service Copenhagen, Denmark Tel: +45 31 18 00 88 Fax: +45 31 18 35 60 Mr. Peter Laursen
-	United Kingdom Ireland	Sheila Pantry Associates Sheffield, United Kingdom Tel: +44 1 909 77 10 24 Fax: +44 1 909 77 28 29 Mrs. Sheila Pantry
-	Netherlands	Netherlands Institute for the Working Environment, NIA Amsterdam, the Netherlands Tel: +31 20 549 84 77 Fax: +31 20 644 14 50 Mrs. Sonja Nossent
-	Europe in general	Eurostat Luxembourg, Luxembourg Tel: +352 43 01 34 9 96 Fax: +352 43 01 34 4 15 Mr. Johnny Dyreborg
		European Foundation for the Improvement of Living and Working Conditions Dublin, Ireland Tel: +353 1 28 26 888 Fax: +353 1 28 26 456 Mr. Henrik Litske

The following institutes and persons contributed to, or largely compiled, the national report on their country:

- Austria

  Ing. Langer
  AUVA
  Adalbert-Stifter-Strasse 65
  A-1200 Wien

  Mr. Herbert Krämer
  Kammer für Arbeiter und Angestellte für Wien
  Prinz-Eugen-Strasse 20-22
  A-1041 Wien

  Dr. Franz
  Osterreichisches Statistisches Zentralamt
  Abteilung 5: Sozialstatistik
  Hintere Zollamtstrasse 2 b
  A-1033 Wien

  Dr. Michaela Moritz
  Ostereichisches Bundesinstitut für Gesundheitswesen
  Stubenring 6
  A-1010 Wien

- Belgium

  Mr. de Brabander
  Nationaal Instituut voor de Statistiek (NIS)
  Leuvenseweg 44
  1000 Brussel

  Mr. de Lange
  Verbond der Belgische Ondernemingen (VBO)
  Ravenstein 4
  1000 Brussel

  Mr. Goffinghs
  Ministerie van Tewerkstelling en Arbeid
  Administratie van de Arbeidsveiligheid
  Belliardstraat 51
  1040 Brussel

  Mr. Steen
  Minister van Tewerkstelling en Arbeid
  Administratie van de Arbeidshygiëne en -geneeskunde
  Belliardstraat 51
  1040 Brussel

  Mr. Theunissen
  Fonds voor Arbeidsongevallen (FAO)
  Troonstraat 100
  1050 Brussel

Mr. van Assche
Fonds voor de Beroepsziekten (FBZ)
Sterrenkundelaan 1
1050 Brussel

- Finland
Mr. Timo Kauppinen
Työterveyslaitos
Tel: +358 0 47471
Fax: +358 0 414634

- France
Dr. M. Falcy
Institut National de Recherche et de Sécurité (I.N.R.S.)
Rue Olivier-Noyer 30
75680 Paris, Cédex 14

- Germany
Dr. Karl Kuhn
Bundesanstalt für Arbeitsschutz
Friedrich-Henkel-Weg 1-25
D-44149 Dortmund

Mrs. Christel Streffer
Bundesministerium für Arbeit
Rochusstrasse 1
D-53123 Bonn

- Luxembourg
Mr. Demuth
Association d'Assurances contre les accidents
Route d'Esch 125
L-2976 Luxembourg

- Sweden
Ms. Madelaine Bastin
Statistiska Centralbyran
Tel: +46 8 783 4645
Fax: +46 8 783 4916

Annex 2 Contents of the matrix and guidelines for national data gathering

Contents of the matrix

1. Description of the context
   - 1.1 Labour market information: national labour force
   - 1.2 Labour market information: companies in the country
   - 1.3 Organizational context
   - 1.4 Information regarding the context at sectorial level

2. Description of the working environment (see Annex 5 for further specifications)
   - 2.1 Physical exposure
   - 2.2 Chemical exposure
   - 2.3 Biological exposure
   - 2.4 Psycho-social exposure (organizational constraints)
   - 2.5 Physiological exposure
   - 2.6 Exposure to safety risks
   - 2.7 Other exposures
   - 2.8 Exposure data at sectorial level

3. Health and safety output
   - 3.1 Occupational accidents
   - 3.2 Occupational sickness absenteeism
   - 3.3 Sickness absenteeism in general
   - 3.4 Occupational morbidity
   - 3.5 General morbidity
   - 3.6 Reported occupational diseases
   - 3.7 Occupational disablement
   - 3.8 General disablement
   - 3.9 Occupational mortality
   - 3.10 General mortality
   - 3.11 Health and safety output data at sectorial level

4. Trends and strategies regarding data production
   - 4.1 Description of trends and strategies

Guidelines to the matrix

1. At 'Remarks/discussion' under each table in the matrix was asked for clear specifications which would help to understand the national data, and to compare it with data from other countries. The remarks/discussion should contain:
   - the (national) definitions on which the national data are based (what is being included, what is not?);
   - a clear specification if categories, different from the provided ones, were used;
   - explanatory specification on the, perhaps several, sources of the data (authority of sources, whether they concern objective and/or subjective reports (e.g. on perceived exposure), the methodology for data gathering in this source etc.);

   Further remarks/discussion may concern:
   - explanations or background information on the issue described in the table (what might seem self-evident in one country is not in another);
   - important developments and trends regarding the issue;

- explanations on the lack of data or the lack of quality of data (for example due to underreporting or underregistration);
- interpretation of the data (for example explicit notes on possible subjectivity);
- in section 2 and 3: additional data on specific risk groups, such as interim workers, student workers etc.

2. To collect the national data, the following sources of information have been suggested:
   - national statistics;
   - national registers on:
     * occupational accidents;
     * sickness absenteeism (occupational and non-occupational);
     * reported occupational diseases;
     * recognized occupational diseases;
     * morbidity (occupational and non-occupational);
     * mortality (occupational and non-occupational);
     * chemicals and products;
     * workplace measurements;
   - national and international surveys, such as:
     * national labour force survey;
     * national surveys on work environment;
     * Eurostat labour force survey;
     * the First European Survey on Work Environment 1991-1992 (European Foundation, 1992);
     * the HASTE Database' and monitoring systems mentioned therein (European Foundation for the Improvement of Living and Working Conditions (1994b);
   - international statistics, such as:
     * ILO annual reports;
     * WHO annual reports;
   - other authoritative national sources;
   - less authorirative, but professional literature;
   - key informants, in which case the following procedure has been recommended:
     * firstly to collect as much data as possible from the other sources;
     * to take up these data in the matrix;
     * if supplementing the data was considered necessary, to discuss the paper with the chosen key informant(s). The supplementary information may for example have consisted of: additional sources, estimations of exposure groups or incidence of diseases, accidents etc.

3. Explicitly was asked to always mention the reference period and the specific information source from which the data stem.

Annex 3 International Standard Classification of Occupation ISCO-88 (COM)

The two-digit level is as indicated below (Eurostat, 1992b):

0 Armed forces:
- 01 Armed forces

1 Legislators, senior officials and managers:
- 11 Legislators and senior officials
- 12 Corporate managers
- 13 Managers of small enterprises

2 Professionals:
- 21 Physical, mathematical and engineering science professionals
- 22 Life science and health professionals
- 23 Teaching professionals
- 24 Other professionals

3 Technicians and associate professionals:
- 31 Physical and engineering science associate professionals
- 32 Life science and health associate professionals
- 33 Teaching associate professionals
- 34 Other associate professionals

4 Clerks:
- 41 Office clerks
- 42 Customer services clerks

5 Service workers and shop and market sales workers:
- 51 Personal and protective services workers
- 52 Models, salespersons and demonstrators

6 Skilled agricultural and fishery workers:
- 61 Skilled agricultural and fishery workers

7 Craft and related trades workers:
- 71 Extraction and building trades workers
- 72 Metal, machinery and related trades workers
- 73 Precision, handicraft, craft printing and related trades workers
- 74 Other craft and related trades workers

8 Plant and machine operators and assemblers:
- 81 Stationary-plant and related operators
- 82 Machine operators and assemblers
- 83 Drivers and mobile plant operators

9 Elementary occupations:
- 91 Sales and services elementary occupations
- 92 Agricultural, fishery and related labourers
- 93 Labourers in mining, construction, manufacturing and transport

Annex 4 *Statistical Classification of Economic Activity in the European Community NACE-1970*

The one-digit level of this former version of NACE (1970) consists of the following ten main sectors (European Foundation, 1992; Annex 3, p. 227):

0 Agriculture, forestry and fishing

1 Energy and water

2 Extraction and processing of non-energy producing minerals, chemical industries

3 Metal manufacturing, mechanical and electrical industry

4 Other manufacturing industries

5 Building and civil engineering

6 Distributive trades, hotels, catering, repairs

7 Transport and communications

8 Banking and finance, insurance, business services, renting

9 Other services

This classification systems has been revised in 1993 (see NACE Rev.1, the cleaned-up version of the version in the Official Journal of the EC L 83 April 1993, by Eurostat, Luxembourg, June 1993).

It was chosen to use the NACE version of 1970 however, because it was likely that most of the desired data (preferably from 1993 or 1992) would be available according to that classification.

Annex 5 Exposure Classification System - a Proposal

The list is mainly based on a proposed exposure classification system, which was used by DG V of the European Commission in a project on occupational diseases (P. Laursen et al, 1994):

1 Physical exposure:
- 11 Noise (harmful, annoying noise etc.)
- 12 Vibrations (hand-arm, whole body vibrations etc.)
- 13 Climate/thermal conditions (temperature, draught, humidity, indoor/outdoor etc.)
- 14 Radiation (ionizing, non-ionizing, optical radiation)
- 15 Lighting/conditions of sight (lack of daylight, reflections, blinding etc.)
- 16 Pressure (positive, negative pressure)

2 Chemical exposure:
- 21 Specific chemical agents
- 22 Materials and compounds
- 23 Products
- 24 Dust, vapour, fumes

3 Biological exposure:
- 31 Viruses
- 32 Bacteria
- 33 Growths (fungi, mosses, plants)
- 34 Animals (protozoa, worms, insects, fish, birds, mammals etc.)
- 35 Foodstuff (animal, vegetable origin)
- 36 Vegetable fibers, wood, wood products (including dust)
- 37 Other materials from biological origin (skin, furs, seeds, bones etc.)
- 38 Biochemical material (enzymes, proteins etc.)
- 39 Material from human beings (tissues, blood etc.)

4 Psychosocial exposure (a.o. organizational constraints):
- 41 Working pace (hectic, machine-controlled pace, time constraints etc.)
- 42 Form of payment (payment related to production, risks, working hours etc.)
- 43 Job content (dull work, monotonous work, high requirements, large volume etc.)
- 44 Working hours (shift work, fixed night/evening work, inconvenient/irregular hours, overtime, long working hours ($\geq$ 10 hrs/day, $>$ 45 hrs/week) etc.)
- 45 Influence and control over own work (influence on planning, execution, organization of work, influence on work method and speed, availability of information, provisions for consultation and participation etc.)
- 46 Social interaction (work in isolation, work-induced or informal social contacts, work support from colleagues or superiors, interpersonal problems)
- 47 Traumatic experience (fear of, or actually experienced violence, sexual harassment, discrimination, accidents etc.)

5 Physiological exposure:
- 51 Working posture and movements (sitting, standing, walking, bending, reaching, turning, crawling too long/often, repetitive work, sudden or wrong movements etc.)
- 52 Manual lifting, handling, pushing and pulling (too heavy objects, heavy work etc.)
- 53 Use and handling of handtools and equipment (too heavy tools, repetitive, hastened or one-sided use etc.)

6 Exposure to safety risks:

- 61 Falling to lower level (by stumbling, slipping, jumping etc.)
- 62 Falling at (almost) the same level (by stumbling, slipping etc.)
- 63 Getting struck by moving objects (falling, swaying, shoving objects, stubbing and bumping up to machines etc.)
- 64 Getting trapped/pinched/crushed (by moving machine parts or objects, while piling up etc.)
- 65 Getting cut/stabbed (by sharp and pointed objects, including portable implements and utensils)
- 66 Horizontal en vertical transport (tractors, trucks, shovels, lorries, tackles, fork lift trucks etc.)
- 67 Fire and explosions
- 68 Electroshocks and electrocution
- 69 Contact with hot or cold objects (burning, freezing etc.)

7 Other exposures

Annex 6 Diagnostic Categories of Diseases

This list almost entirely corresponds with the three-digit level of the International Classification of Diseases ICD, version 10 (WHO, 1994):

110 Cancer
120 Cardiovascular disorders
130 Dental disorders
140 Eye disorders
150 Gastrointestinal disorders
160 Hematological disorders
170 Hearing disorders
180 Hepatic disorders
190 Irritant effects of the skin or mucous membranes, including disorders of allergic nature
200 Neurological disorders
210 Pulmonary disorders, including disorders of allergic nature
220 Musculo-skeletal disorders
230 Infectious diseases
240 Psychological disorders
800 Accidents
998 Diagnosis not elsewhere mentioned (please specify)
999 Diagnosis unknown

Annex 7 Overview on the total working population in 16 European countries

Country	Year	Total working population*	National population**
Austria(A)	1985	3,234,500	7,555,338
	1993	3,608,100	7,907,600
Belgium(B)	1991	3,730,500	9,987,000
	1993	2,532,719[1]	
	1993	3,745,500[1]	10,068,300
Denmark(DK)	1990	2,670,000	5,135,400
	1992	2,609,859	5,162,100
Spain(E)	1992	12,366,000	39.055,900
	1993	8,686,000[2]	39,114,200
		11,838,000[2]	
Finland(FIN)	1990	2,488,000	4,974,400
	1992	2,199,000	5,029,000
	1993	2,064,000	5,055,000
France(F)	1989	22,146,000	56,017,000
	1992	14,440,400[3]	57,217,600
Germany(GER)	1991	37,445,000[4]	79,753,200
	1992	36,940,000[5]	80,274,600
Greece(GR)	1990	3,719,100	10,046,000
	1991	3,632,400	10,120,000
Ireland(IRL)	1993	1,125,000[5]	3,560,000
Italy(I)	1989	21,154,000	57,504,700
	1991	21,595,000	57,746,200
	1992	21,609,000	56,757,200
Luxembourg(L)	1992	196,800	389,800
	1993		395,200
the Netherlands(NL)	1989	6,155,000	14,805,200
	1990	5,885,000	14,892,600
	1993	5,925,000[6]	15,239,200
Norway(N)	1993	2,004,000	4,299,200
Portugal (P)	1993	4,457,600	9,900,000
Sweden(S)	1993	3,964,000	8,692,000
the United Kingdom(UK)	1993	25,629,100	57,959,000

Sources:

*) International Labour Office (ILO), Yearbook of Labour statistics 1994[b], p. 237-242 (employment), Geneva 1994.

**) Eurostat, Basisstatistieken van de Gemeenschap 1994, p. 115 (national population), 31st edition, Luxemburg 1994.

Remarks:

1) Belgium: the percentages are based on a total working population of 2,532,719 employees in 1993 (excluding 600,000 employees from the governmental and educational sector). The second number for the total working population in 1993 was extracted from the ILO-Yearbook 1994.

2) Spain: the first number refers to a total working population that excludes the self-employed workers. Due to the fact that a lot of data was extracted from a survey that only included the employed workers, this number was used to calculated the percentages.
The second number refers to the total working population (including the self-employed).

3) France: this number was given in the French report. The ILO-statistics on the total working population of France in 1992 is 22,285,000.

4) Germany: the total working population of West-Germany (29,684,000) and East-Germany (7,761,000) were taken together.

5) Ireland: there were no data on the total working population of 1993. This number refers to the total working population of 1992.

6) Netherlands: the total working population of 1993 consist of employed and self-employed workers who work more than 12 hours a week.

Annex 8 Age categories used in the matrix

In the matrix the following age categories were provided to obtain data according to age:

1. $\leq$ 13 years
2. 14-24 years
3. 25-49 years
4. 50-64 years
5. $\geq$ 65 years

If national data according to age were available, but not corresponding to these specific categories, authors were requested to provide these available data and to specify the categories they had used.

European Foundation for the Improvement of Living and Working Conditions

EUROPEAN WORKING ENVIRONMENT IN FIGURES
Availability and quality of occupational health and safety data in sixteen European countries

Luxembourg: Office for Official Publications of the European Communities

1996 – 162 pp. – 21 x 29.7 cm

ISBN 92-827-6552-0

Price (excluding VAT) in Luxembourg: ECU 16.50